T0247888

THE QUIET MOON

Praise for *The Quiet Moon*

'Lunar curiosity guides this graceful sharing, not only of Parr's deep knowing of the natural world, but also his vulnerability as he wrestles with mental health. A beautifully personal book that's bound by a sweet melancholy.'

Verity Sharp, broadcaster

'Kev Parr stands among the finest natural history writers of our generation … a masterful storyteller and a man wholly unafraid to bare his soul on the page. With him, the reader is blessed with the most thoughtful guide and companion.'

Will Millard, BBC presenter and author of *The Old Man and the Sand Eel*

THE QUIET MOON

Pathways to an Ancient Way of Being

KEVIN PARR

FLINT

To Zac, Millie, Ollie, Bertie, Erin, Bade and Rufus.

Cover image: Andy Lovell (andylovell.co.uk/)
Internal images: Sue Parr

First published 2023

FLINT is an imprint of The History Press
97 St George's Place, Cheltenham,
Gloucestershire, GL50 3QB
www.flintbooks.co.uk

British Library Cataloguing in Publication Data.
A catalogue record for this book is available from the British Library.

ISBN 978 0 7509 9869 7

Typesetting and origination by The History Press
Printed and bound in Great Britain by TJ Books Limited, Padstow, Cornwall.

Trees for Life

CONTENTS

PROLOGUE

A GLINT OF MOONLIGHT

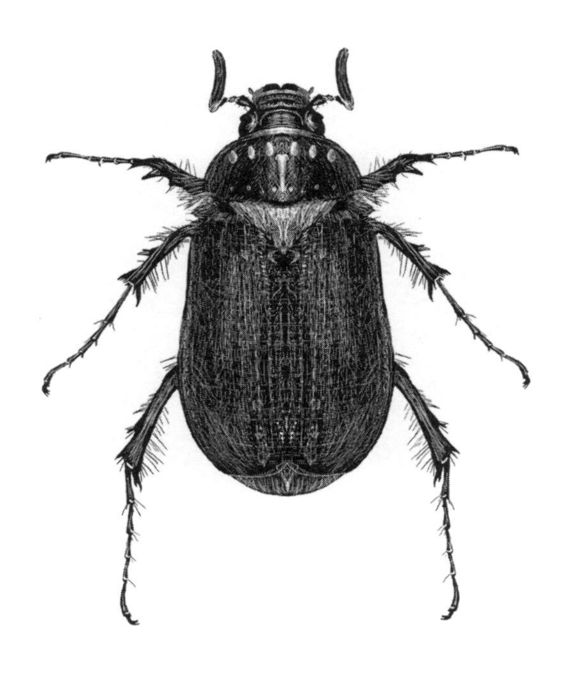

The owl of Minerva spreads its wings only with the falling of dusk.

(Georg Wilhelm Friedrich Hegel, *The Philosophy of Right*)

THE SUN HANGS for a final moment. Bloodied like a bruised orange, but sharp-edged and spent. I can look straight at it, the dying ember of a flaming day. Its energy remains though. The still that surrounds me remains thick with the heat. Not a breath of wind, not even here where there is always a blow.

A bead of sweat drops from my eyebrow on to my cheek, smearing across the lens of my glasses as it falls. Suddenly I am aware of the damp around the neck roll of my T-shirt and the small of my back. 'It's good exercise,' I remind myself, reaching for a tissue from my pocket to buff up my glasses. I'd almost swapped my shorts for long trousers before I left the cottage, but this evening is one of those fitful high-summer nights when the whole world has to stop.

There it goes. As it meets the horizon, so the liquid of the sun begins to ooze like the yolk of a perfectly poached egg. Smearing into the silhouette of Lewesdon Hill and then spilling in a lava bubble that sparks and fizzes and forces me to squint and blink my gaze away. A glorious sunset, although not as blazing as some. The haze and low cloud have diffused the dusk colours, softening the yellows and golds into mauve. It feels appropriate to the mood, though. The thick air that has plumed from the Sahara has brought a bout of brutal heat – even here in the West Dorset rolls where the fresh of the sea usually keeps the temperature in check. But I like it. I like the unusualness. The sun has all but set and I am feeling over-dressed in a T-shirt and shorts. There is a flavour of the Mediterranean in the air but with a serving of local seasonal vegetables. And the smell of this evening is the thing that I am finding most extraordinary.

Around my feet is a carpet of colour unlike anything I have seen on Eggardon Hill. There are harebells and bird's-foot trefoil, red clover and lady's bedstraw. There are thistles and grasses and a multitude of other plants that I cannot name. As I first stepped through the gate and on to the fort, and the perfume tickled my nostrils, I bent down among it all to try to find the actual source. There was no singularly distinctive waft, something powerful like a honeysuckle or lime, and none of the flowers or seeding grasses that I put my nose to seemed to have any great scent of their own. Instead, I was smelling everything – from the pollen and nectar down through the sun-warmed leaves and the exposed soil and desiccated sheep-shit. It is power in numbers. All of the most subtle odours teased by the heat and then allowed to simply hang. With no breeze or coolness to dissipate, I was probably even catching the whiff of the miniscule eggs that the marbled whites were scattering. All of these things coming together to create one glorious infusion. I was smelling the hill itself, the millennia of change in geology and ecology that had led to now. And even as I looked west at the setting sun, I couldn't help but be distracted by the unexpected intensity that was provoking a different sense.

I hadn't come up here to smell the air. I came alone but wanted to share the sunset. Or rather, I wanted to look at the sunset as others might once have done. The people who first took tools to the earth of this hill over 5,000 years ago, building ditches and ramparts, creating sanctuary in altitude, structure among the wild. What must they have thought to watch the sun disappear only to leave a trail of colour in its wake? Would they have cared? They would surely have stood as I do now, drawn unconsciously towards the fading warmth. Not least because of the connection with Lewesdon Hill and Pilsdon Pen that stand in the west. Both were topped with forts such as this one, as so many of the hills in Dorset are. Perhaps they would see the dance of flames from the homes of their counterparts. Fires lit to cook and communicate. A sense of visual connection they shared with the light of the sun itself.

I smile. I like the thought of standing where those people stood. At moments such as this I feel my own connection with them. An appreciation of a moment – this moment – and I don't doubt that on an evening such as this they would have gathered together and simply watched. The world would have looked very different then, of course. And even as the hedgerows, field systems and electricity pylons dissolve into the dark, the lights of houses, the town of Bridport, a distant but arresting red-lighted mast, all remind me of the impact that humankind has had on the entire landscape. The wildlife would have been different too, but it is difficult to determine to what extent. There would have been some agriculture, but there were no pesticides, fungicides or combine harvesters. There was also but a fraction of the people. It is hard to visualise to what extent this part of Dorset would have been moulded by the hands of humans, when all around me today is managed.

The grass around my feet thrums with the chirrup of grass-hoppers, while far below me, where the grass slips into a pocket of scrub and trees, comes the soft coo of wood pigeon. The colour of the trees grabs my interest – or rather, the lack of colour. If I move my eyes quickly across the tangle of treetops I can make out the green, but as I stare still so the colour vanishes in the lowering light and the shapes roll and swirl as my eyes try to fill in the gaps between what they know is there and what they can actually discern. The rods in my eyes stirring, only for the cones to dance as I look back up at the technicoloured horizon. There, the orange and gold of the sunset are already climbing higher and spreading. As the earth slowly turns so the sun finds more atmosphere to shine through. It has vanished from view and yet for a time it will brighten even more of the sky.

Dzzzzzz. What's that? A faint insect buzz has quickly loudened and the source is now donking me on the head. I don't recognise the sound. It isn't a bee, wasp, hornet – a beetle perhaps? Ah! It's a chafer. Not one of the big, fat May-bugs but a smaller summer cousin. It certainly seems to find me rather interesting, in particular the top of my head. There is no aggression to its behaviour but the sweat on my forehead is quite a draw, and a second chafer has now joined the first. Time to wander, perhaps. A slow mooch back around the southern slopes.

I glance back at the western sky, the mauves deepening near the horizon as the oranges lose their vibrance but stretch ever higher into the night. I still feel the pull – an urge to follow the colour and light. And the contrast as I turn away is marked. The eastern sky is deep blue and starless, the grasses and flowers of the ridge behind me all faded to grey. There is a sense of nothing about it, an emptiness that couldn't contrast more sharply with the view west. But as I begin to tread a watchful path back, my route edges me out from beneath the loom of the inner rampart, and there, sharp and silent, is the moon.

It is surprisingly white, given the thickness of the air, although it is already quite high in the sky so has little atmosphere through which to yellow. I'm not sure quite when the moon will next be full, but it is waxing towards it, and judging by its size, is around four days away. The Moon of Claiming, a period apparently when the male Celts would venture out to 'claim' their wives. A primitive notion to say the least, although the more I learn about the people who would have built this hill fort, the more I question that sort of presumption. Perhaps I should rephrase that sentence – it is less the things I have learned and more the things I have unlearned. That touch of Roman propaganda that we have so long taken as read. And much as I, a white British man, have recently become aware of the way in which the history I have been taught has been delivered to suit my own demographic, so I realise that all history should be questioned to some degree.

More pertinent, though, is less what it represents and may have represented at this precise point of the year, and more how it reflects a pattern of process that I have subconsciously adopted. A month ago, the sun would have set to the right of Lewesdon, and in another month's time it will be much further south, to the left of Pilsdon Pen. A natural calendar of sorts, although only precise if you line two points exactly together. The moon, on the other hand, will tell you the day no matter where it sits in the sky. A practised eye would recognise how far into the current quarter the moon has stretched: it would be almost as straightforward as flicking on a mobile phone or glancing at a wristwatch to check the date. And it wouldn't matter that the moon is sometimes obscured by cloud because once you learn to trust your subconscious clock you soon realise how reliable it can be.

The key, though, is to allow yourself the opportunity not to care. It isn't easy, much as if someone were to tell you to clear your mind of all thought and your head immediately floods with a sense of everything. To be able to lose yourself in a moment is only possible if you are able to not try to achieve that very thing. Rather like that moment when drifting off to sleep, when your thoughts start to drift sideways as your subconscious begins to whirr. As soon as you consider what is happening, you lift yourself straight back to wake.

Time is a harness because of our interpretation of its passing. Because we have to be somewhere or meet someone. Because we create goals for ourselves that relate to certain stages of our lives. Our expectations can be driven by human instinct; the pains that Sue, my wife, feels about us not being parents are compounded by a body in grief. The dreams I have where a child is mine, is ours, and yet is somehow out of my reach do not need Freudian analysis. And certainly I did not expect to nudge into my late 40s and find myself childless, renting a house and without savings. Wearing clothes with more holes than fabric and relying on my parents for grocery drops or help for when the starter motor on the car packs up.

Sometimes we are forced to walk paths that we don't want to take, yet there remain the assumptions of what will come. Fifteen years ago, after a fairly carefree meander, the pieces had pretty much slotted in place. A mortgage on a new build, enough income to be putting several hundred pounds away each month. The future was ours to make and at a pace of our choosing. Even when the threads began to unravel there was no immediate panic. Sue would get better; I would write a best-seller. Selling up was a short-term necessity but it was *short term*. And there were always going to be children at some point; that was something so inevitable that it didn't need to be discussed. Which then, of course, makes it all the more painful when reality finally rattles at the door. Time catches up with us all.

The compulsion that found me on Eggardon Hill this evening is linked to all of those things. But only because I saw the date and associated it with a ten-year anniversary. It is a decade since we moved to Dorset and that caught me a little off guard. To the extent that I created my very own tumult. A whirl of regret, shame and disappointment. I was a 13-year-old boy again, falsely believing that school marks were all my parents and teachers cared about. And then came the glorious grounding. The smell and the sunset. A reassurance of all that we do have – and that is plenty.

I walk a few steps and then pause again, looking once more upon the moon. As I do, my eyes catch a movement of light further away and below the horizon. A double-decker bus, probably a late-running Jurassic Coast Special, is trundling westward along the A35, around 2 miles away. I can hear the faint hum of the road when I listen for it, but I had detached from it in order to let the grasshoppers and chafers flood my ears. There I was, a moment before, perhaps not quite imagining myself as an ancient Celt, but certainly feeling some sort of connection to that time. Seeing the moon and perhaps sharing a similar understanding as to the concept of time. And, there in the distance, a reminder of today. Of modern life. Two rows of bus windows lit up and moving at speed. It is far too distant to determine if anyone is behind those windows, but rather like a train passing at night, I have a snapshot into another world lived at a different pace and no one there knows that I am here watching. It reminds me a little of standing on a bridge and looking into a stream below. The constant yet ever-changing movement of water that is on such a different course to my own. Two paths cross but do not, and cannot, mingle in that moment. So separate are they of purpose.

I reach down to feel the damp of the grass but it is slight. The heat and dusty air holding back the dew. Still I tread carefully as I make my way down the southern slope, aware that even a light moistening could make for greasy footfall, and as I do I notice my shadow. The moon is bright enough and high enough in the blackness to cast my form and in a moment my presence seems more tangible. As I had slipped out of my own mental constraints, so I had become less aware of my physical being. Now, though, my moonlit shadow moves slowly with me, reminding me of myself. And that self is so much more content than it had been when I arrived. The anxieties and self-deprecation melting into the mauve. All it took was for me to step sideways for a moment and to forget how long a moment should actually be. It doesn't matter how old I am, or what I haven't got or achieved. I am gloriously insignificant. Almost as though coming here tonight on a different pretence was actually my subconscious pulling together the things that I needed to ground again. To step away from Time.

I have learned a lot this year and hope to continue to do so. But perhaps the most important lesson is here right now. It is high summer, but in a couple of pages' time I will take you back to the cold and dark of midwinter, because that is where we always start. But the sun doesn't rise as the clocks strike twelve on New Year's Eve and nor does the moon sit full. The cycles are ongoing, regardless of how much we try to shepherd them. Sometimes we need to break our own rules and routine to remind ourselves of that. After all, the Celts didn't begin their year with fireworks and *Auld Lang Syne*, if they even began a year at all.

And that is why I can write an introduction to a book that is already halfway written. It's just taken me six lunar cycles to realise the fact.

1

THE QUIET MOON

An absolute silence leads to sadness. It offers an image of death.

(Jean-Jacques Rousseau, *The Reveries of the Solitary Walker*)

IT'S GREY AND STILL, one of those winter days when the world forgets to wake up. There have been moments when the sun has threatened an appearance, but the blanket of cloud has remained, the temperature barely shifting from night into day. These are the sorts of conditions that the angler within me longs for at this time of year – mild air and low light levels can make for excellent fishing. But my rods are at home and my body and mind need a different kind of interaction.

Daily exercise had felt more achievable last March when we first went into lockdown. Helped no doubt by a sense of novelty about the situation, but likely too because spring was gaining some momentum and bringing so much possibility to the world. I also had a heightened appreciation of the place where we live. Stepping straight out into a landscape that millions of other people could only dream of as they stared on to concrete from the prison of their own circumstance. I was encouraged to exploit it because I felt such privilege; to not make the most of living in rural isolation would have almost been insulting to those who would have wished to be. And we were all in it together – there was, in the beginning at least, a sense of togetherness as the whole world came to terms with a shared threat. We thought of those living alone and the less able. Those stuck in high rises or separated from loved ones. And we shared what we could – photographs, podcasts, little snippets of birdsong. We Zoomed and quizzed and shared virtual pots of tea. And all the while, the days were lengthening, spring would lead into summer and all would be well.

The circular walk around our own little patch of West Dorset, which I might usually make once a fortnight, became a daily stomp. There was, admittedly, a lengthy pause halfway to scan through the mixed flock of buntings that had settled in number in the stubble and maybe snatch a glimpse of the merlin that shadowed their winter. But I began to lose weight, to breathe deep and almost – almost – feel a bit better about myself.

Ten months on, though, and I have slipped back into familiar habits. The hangover from a cancelled Christmas and New Year has been tempered by a continuation of excess. The weekends well wetted with cheap cider, while a glut of carbohydrates replaces the alcohol during the week. The house is cold, my afternoon naps are increasingly difficult to wake from and I'm finding too many excuses to avoid any kind of physical exertion. And while I know what is good for me, what will actually make me feel sharper and happier, the lack of motivation is coupled with a depressive cycle that has whirred for most of my life. There comes a point when *won't* becomes *can't*, and sometimes *can't* actually feels like a curious sanctuary.

Today, though, I was prompted out for a purpose not my own. A visit to my parents' house to help out with a few urgent chores around the garden. It is always easier to nudge myself out if it feels as though I am obliged for someone else's benefit, a state of mind that will in turn be beneficial for me. The road to Beaminster (where my parents live) is a lovely one, dotted with places to stop on my way home for my daily exercise. I chose this spot because I don't know it very well. It is close to one of my autumn mushroom haunts, where a mossy roadside verge can sparkle with chanterelles. I first stumbled upon it several years ago when exploring some of the quieter lanes of this already quiet area. Taking ever more varied routes from home to the weekly supermarket visit in Bridport, windows down and the car doing a very lazy trundle. I was slipping through a thick beech corridor when a spill of gold caught my eye. A lovely clamber of chanterelles glowing against the green like the first celandines of spring. I was a little bit too excited, though, and didn't consider how deep the mud was in the spot where I pulled over. The nearside wheels sank and the car bottomed out, leaving me marooned and with barely any battery on my phone to call for aid. The tranquillity of that little lane leaving me feeling rather isolated. It took a kindly soul in a Land Rover to drag me back on to the road, and I later delivered a basketful of chanterelles to his door as a thank you.

Visits here have since been fleeting. A quick, early-autumn diversion to check the mushroom spots before journeying on elsewhere. A couple of years ago, though, I came on a whim. It was later in the year and winter was already nibbling at autumn's toes. The beeches had gone to brown and I didn't loiter long beneath them. Instead I pushed through the trees and picked up a footpath that I had seen marked on the Ordnance Survey map. The local slopes and pasture are crisscrossed with public rights of way, and the majority seem never to be trodden. Often it is a case of picking out the footpath signs and forming your own route between them, and, providing the gates are left as they are found and respect is given, there is nothing but a cheery wave should a farmer or landowner appear. On that day, having followed the path I had seen on the map for a time, I picked up another route and trod into open country where I discovered a meadow filled with waxcaps. The dew had added an extra sheen to their glisten and even the white snowy waxcaps had a sugary shine like polished marble. The sight of those fungi had elevated my walk and given me cause to return. Today might be too late in the season to see any mushrooms but I am curious as to what else I might stumble upon. The one issue being my own state of mind. I am not feeling particularly open to opportunity and have a niggle that is urging me to return home and draw the curtains. The hour is already late, so what is the point in lingering?

I take a deep breath and press on to the gate that leads me out into the open country, pepping myself up with each step. There might be a barn owl hunting, or even a hen harrier. I might see a hare or even a wild boar. Come on Kev, at worst you are getting your heart thumping and stretching your legs.

I pause in the waxcap meadow, beginning to begrudgingly feel the benefit of my push. There is nothing but pasture around my feet, the grasses looking tired and yellowed. Dying back in the cold, rather than gnawed by sheep or cattle. A short effort takes me up on to the summit of a knoll, where a single gorse stands leggy and slightly bare. In the south-west there is a slight hint of a glow. A narrow smear of orange-pink that sits like a letterbox at the foot of a great dark door. As I watch, it smudges back behind the grey of the cloud, but that was the first hint of the sun that I have seen for a couple of days. It also suggests that the hour is slightly later than I thought, and I turn to the south-east, expecting to see the moon. I smile at my absent-mindedness. If the sun was unable to poke a route through the cotton-wool veil of cloud then the moon certainly wouldn't be visible. It would be there about now though. I had noticed it quite high in the sky at dusk two days ago, just before the cloud rolled in, so it should still be rising in the daylight. Instead, though, my attention is drawn to another familiar form. Eggardon Hill looks a little disappointing from this angle. My own elevation and the sweep of the rise to my left have allowed it to meld into a landscape it often dominates. The ridged ramparts of the Iron Age hill fort that sits atop it are distinctive, however, and take my thoughts in a different direction.

I remember as a child visiting the Bosworth battlefield. I was around the age of 10, and my family were staying with friends, my godfather's family, in Leicestershire. I knew nothing of the history involved, or the relevance to what little history I knew, but was intrigued as to what I might see. As it turned out, I was wholly underwhelmed. All I was looking at were fields and hedgerows, through the muck of a damp and blustery day. I have no idea what I expected to witness, but given the build-up, the talk of civil war and the death in action of Richard III, I certainly didn't anticipate a view little different to those I saw back at home in Hampshire.

In hindsight, I am quite surprised by my response. My imagination has always been quite easy to rouse, and I had a nervous anticipation of what I might see or how I might feel. Yet I found it difficult to visualise anything beneath the surface. I felt no connection. It could be that I was unerringly tapping into the dispute around the actual battlefield location, but far more likely was the fact that with no remnant of that time, I couldn't fill in the gaps. And having grown up in a heavily managed, if rural, area, I had got used to a landscape without historical context.

The folds of West Dorset are less sanitised, largely because the clefts and combes that have formed across this melting pot of geology are less convenient for arable agriculture – proven, perhaps, by the lynchets that remain stepped into the hillsides. Earthworks created by ancient humankind in an effort to exploit every inch of available soil, now regarded as nothing more than grazing pasture. With less pressure on the land, there remains a greater sense of what once was, and there are also far more clues to the past. Alongside the lynchets are hill forts, tumuli, stone circles and sarsens. Most of these are untouched, like the stone burial chamber around half a mile from our cottage. It appears to be partially collapsed but I have found no record of it ever having been officially opened or excavated. And every year the farmer ploughs around it and the crops grow almost high enough to hide away the stones.

As I walk around such places, invariably with a primary interest in the wildlife rather than the historical context, I cannot help but feel a sociological connection to the land. I am less drawn to the actual history and process than to an empathy with the people who once lived here. It is loose, of course, but tangible, and stems from a decade of living outside the perimeters in which I – we – had always expected to live. We sold our flat in Winchester while we still had some equity to call our own. Illness had forced Sue into redundancy and that meant our income more than halved. A move west had always been mooted, to find a spot midway between our families, but it came under slightly forced circumstances. Not that we minded too much. It was an adventure, of sorts, and certainly a chance for me to tread a different path. But we live in a society that doesn't much care for those on the fringes. When you step out of the flow, whether by choice or from being pushed, it is very difficult to get back in the water. And there comes a point where you begin to realise that there are more important things than getting your feet wet. The rhythm shifts. There is a different perspective.

It is no great surprise to feel a sense of warmth from Eggardon while on a walk I'd frankly rather not be on. It is like seeing a familiar face, a reminder of our home just to the east and the security that comes with that. And it is impossible to look at Eggardon and spend time walking the ramparts and ditches and not imagine it as it once was. Aside from the physical reminders that set it apart from a potential battlefield site in Leicestershire is a deeper, ingrained sense of purpose. I will visit to catch a glimpse of the first wheatear of my spring, or to search for the almost unworldly iridescence of an Adonis blue butterfly in the late-summer sunshine, but there are moments when my mind drifts with a people I know little of. Over time, that sense of connection has spread into the wider area, and I find myself wondering if the sense of escape that I seek in moments of need is similar to the deeper immersion that the Celts must have had within the world. I like to think that they felt themselves part of the landscape and not masters of it. A sense of harmony that reached beyond the land that provided food and shelter and into the sky from where they might have drawn a sense of time.

With no religious or Roman influence, it seems likely that the ancient Celts based their calendar around the lunar cycle. I say 'likely' because there is much we don't know about Celtic life and much that we have previously accepted through the eyes of others. Fake news is not a twenty-first-century phenomenon, although the modern interpretation where facts can be dismissed as such because they don't suit personal agenda perhaps is. But history is always read from a singular perspective which in turn reflects popular culture. Much of what we 'know' today about Richard III, who fell at Bosworth, might be spawned through modern interpretations of William Shakespeare's portrayal. Shakespeare was unlikely to present favourably the man who stood as enemy to Henry VII, particularly while the House of Tudor still ruled, but such depictions become so often repeated that they are accepted as fact. It is a situation even more common in the internet age. As remarkable and accessible as the internet can be to research just about any subject you choose, there is no governance or control over content. Copy writing (and I speak from experience) is often a case of simply finding a relevant source and regurgitating the words in a manner that appears different. A false fact can be innocently, if lazily, repeated and over time it will become more established than the truth itself. Propaganda is ever easier to distribute and individuals ever more easily influenced. And, as a result, any one of us can filter out anything we disagree with and find someone else who shares a specific view.

It is a pattern highlighted by Michael Holding when speaking in relation to sports people taking the knee to highlight racism: 'History is written by the conqueror, not those who are conquered. History is written by the people who do the harm, not by the people who are harmed. We need to go back and teach both sides of history.'

The Celts were a people who made no written record of themselves. Instead, history has portrayed them through the writings of others, chiefly those who were their enemies. Roman and Greek scribes were not often complimentary in their descriptions, presenting the Celts as barbarians, primitive and warlike. Aristotle believed them to be a people ferocious and fearless to the point of irrationality, while Julius Caesar wrote of them as squabbling and aggressively partite, the antithesis of Roman civility. It would seem easy to disregard such information and instead draw only from archaeological evidence, but author and professor Barry Cunliffe counsels against this in *The Ancient Celts*, suggesting that 'to reject such a rich vein of anecdote would be defeatist: it would admit to an inability to treat the sources critically'. Wise words. As always, the answers are somewhere in the grey of the between.

My own readings of Celtic history have come from a somewhat niche interest. Sue and I were sitting one evening, lights off and the curtains drawn, watching the moon rise behind the ridge to the south-east of our cottage. There had been a fair bit of media chat about this particular 'supermoon', and though I can't recall exactly what it was being called, I do remember puzzling over its origins. It was a beaver moon or sturgeon moon or something that didn't have an obvious connection to the folklore of the British Isles. I was reminded of the term 'Indian summer', often quoted when we enjoy a spell of settled, sunny weather in the traditional meteorological autumn. For a long time I had falsely believed it to be attributed to the short dusks and heavy dews familiar on the subcontinent, rather than a provenance possibly inspired by Native American culture. It transpired that the names for the full moons were similarly sourced, a reflection of the cultural influence that the New World has had in Britain, and elsewhere in Europe, for several centuries. I found it odd at first. Surely the moon held greater significance to the ancient cultures that once lived here? And yet, it made complete sense. Nothing was written before the Romans arrived so there was simply nothing recorded. And a delve into the chaos of the internet took me deep into a rabbit warren where every tunnel seemed to bring me back to the surface. I could find plenty of patterns, plenty of lists, but the common theme between them, aside from the names of the moons themselves, was the lack of a source. It became a mild obsession, fuelled every twenty-eight days when the moon filled once more and my sleep became fitful. I would invariably follow the same fibreoptic paths and end up at a similarly blank conclusion. Whole days were spent orbiting a subject that had nothing to do with what I should have been doing, until I realised that I was overlooking a rather vital aspect. I was looking for a nice, neat list of twelve names. One for each moon and twelve for the year. Except, of course, that a year is measured by the length of time it takes the earth to orbit the sun. The moon has no involvement in that. So I was effectively looking for something that couldn't exist. This realisation left me with a couple of thoughts. Firstly, this was all getting beyond the capabilities of my mind and my head was beginning to hurt. And secondly, aided no doubt by the first point, how much did it really matter?

I've changed direction, heading north to skirt around the next knoll and then aiming for a gateway where I can pick up the road and have an easy stroll back to the car. I had planned to follow a route down the main slope of pasture and scrub to where the land suddenly buffers and forms a wide, flat area. I stumbled upon it on my previous visit and almost lost a welly as I squelched into the edge. There was a small area of standing water on one side, thick with alder, but probably not a permanent pond. I unwittingly spooked a couple of mallards who poked their way out through the trailing branches and quacked into panicked flight. I watched them circle a couple of times, assured them I wouldn't disturb them any further, and then worked back up on to more solid ground. As I looked back at the chive-green clumps of clubrush, I wondered what might be tucked up out of sight. Snipe were an obvious possibility, hunkered down with long beaks lowered. Eyes aware but stock-still in mottled brown. It was the possible glimpse of a snipe that had led me back in the same direction, but guilt and apathy have combined to prompt this change of course. While I don't need much persuading in this mindset, on this occasion it seems more than reasonable. Snipe are birds rarely seen in these parts unless flushed, and it doesn't seem fair to do so just for the satisfaction I will get from that flash of zigzag flight.

There is a spot closer to home which is reliable for adders but is also a winter haunt for snipe. It is clay-clogged and boggy, and not where you might expect to find reptiles, but here and there are small, solid, brambled islands where the adders find drier ground. One spot offers an almost guaranteed sighting on a sunny mid-February day, but it is impossible to reach without startling a snipe or two, and in recent winters I have not had the heart to look at all. Leave them be, if they are there, and enjoy an encounter that is altogether unexpected. Especially if it means I can go back to the car, go home and find my own patch of rushes to bury myself in.

I climb the gate rather than open it, a habit that has formed in less than a year. Being closed off from a world suddenly obsessed with sanitisation can create some interesting new obstacles when you do venture out. The latch of a gate – who might have touched it and how long ago? Perhaps a wipe with a tissue will make it safe or working the mechanism through the sleeve of my coat. Is this verging on hysteria? Perhaps if I vault the gate then I can negate the whole issue. It is curious how quickly quirks in behaviour become the norm during a pandemic, although I do climb my gates with care. I once managed to break the bottom rung of a wooden fence with a slightly heavy-footed clomp. I was able to prop it up but had a guilt-filled night and returned the next day with a hammer and splice to make a better repair.

The unforgiving tarmac is not a pleasant surface to walk upon, especially in cheap wellingtons, but the route back to the car is a steady climb that I can push along and get my heart pumping. A compromise to my inner critic for cutting my walk short. As I reach the trees, the air thickens a touch and I catch the waft of pine. Just a hint, not enough to sink into and lose myself in, but it should sharpen as I curve into the main body of forest where the conifers outnumber the beech and hazel. As the lane begins to turn, though, it is another smell that hits me and I break stride in response. The musk of fox hangs heavy, and I slow my breaths to avoid too deep an inhalation. It is a pungent, sulphurous odour that seems to intensify when it gets inadvertently trodden into a carpet or rolled deep into the matted fur of a pet dog. It seems less offensive outside though, when tempered by the damp air, and the slight floral edge can make it mildly evocative. The scent is supposedly comparable to that of a violet, and the supracaudal gland that produces it is often named accordingly. Other mammals, including dogs and badgers, have a similar gland, but the sweetness seems to be especially distinctive to foxes.

I glance around, half expecting to see the animal itself, but sights of foxes are normally fleeting and distant. An after-dark drive around the lanes might offer a glimpse in the headlights and the bitches bolden when they have a den of hungry cubs to feed, but the fox is a mammal more often heard than seen. Especially in January.

A fox courtship is a short, intense love affair. The vixen is in season for a matter of days and throughout the preceding fortnight her beau will not leave her side. They will hunt together, sleep together and mate frequently in order to ensure fertilisation occurs during that most brief of oestrus cycles. The act of copulation might appear painful, and many people believe it is that moment when the vixen cries her pained torment. Others, though, believe that she screams in order to attract a suitor and make herself known, while further study has suggested that both the dog and vixen produce the wail perhaps as a declaration of amour. Whatever the precise reasoning, there are many of us who have sat bolt upright in bed having been shocked awake by a scream that is truly blood-curdling.

The eighteenth-century French naturalist Georges-Louis Leclerc likened the sound to that of a peacock, but it can be both otherworldly and eerily human as it slices through an otherwise silent winter night.

Humankind has long had an uneasy relationship with the fox. We have persecuted him for stealing hens and raiding pheasant pens and yet we encourage him by occasionally leaving the coop door open and releasing millions of part-domesticated game birds every year. We leave rubbish bags in the street and we drop litter by the ton but we bemoan the foxes for making meals of our waste.

So affected is he by human influence that the urban fox has evolved with very different characteristics from his rural brother. He is bold and fearless, and happily diurnal. His bright red fur and thickset brush have become grey and spindly as he trots through dusty pavements and leaps garden fences. I have watched an adult fox slink down Richmond High Street on a warm summer afternoon. He was utterly unfazed by the crowds of shoppers and interestingly no one seemed to notice him. Perhaps in the eyes of some he had simply become another homeless person on the street, ignored, as many are, to the point of invisibility.

In his more traditional habitat, the fox is still considered by many as vermin and he maintains a low profile as a result. He is nervy, still hunts as much as he scavenges, and keeps less sociable hours. Daylight glimpses are less usual and give the impression of an animal far scarcer than he actually is. But there is a fascination to be found in observing a truly rural fox. An unexpected encounter still delivers a shot of adrenaline. He is, after all, a hunter and natural competitor. And although he will run, we will always watch him out of sight, as if he is likely to creep back behind us and play a nasty trick when we are not looking.

Humankind has long tagged the fox as a creature of cunning who is not to be trusted. In ancient Greece, Aesop wrote many of his fables with a fox as a central character. The fox is portrayed as a clever animal, certainly when he outwits the sick lion, but also as one who charms for his own benefit and who can dish plenty out but not be so happy to take it. Aesop seems to regard the fox with a grudging respect but ensures he always gets his comeuppance.

The image of a sly, cunning trickster is a moniker that the fox wears to this day. In medieval Europe the character of Reynard the Fox evolved, sparking adaptations within French, Dutch and German folklore. In each culture Reynard was anthropomorphic and always seemed to walk the less respected paths of society. Geoffrey Chaucer penned his own nod to Reynard in one of his *Canterbury Tales* of the 1390s. There is some mystery to the true origin of the storyline, but *The Nun's Priest's Tale* tells the story of Chauntecleer, a cock whose crowing was unequalled throughout the land.

Chauntecleer dreams of his own demise at the hands of a fox, the same animal that had previously tricked his mother and father. Despite his wife's assurances to the contrary, the dream does indeed come true and Chauntecleer comes face to face with his subconscious tormentor. The fox is wily and charms the cock into crowing, at which point his eyes are shut and he is easy to grab. With most of the farmyard in hot pursuit, Chauntecleer then displays cunning of his own and persuades the fox to stop running in order to taunt the chasing throng. The fox does so, but as soon as he opens his mouth, Chauntecleer flies out and finds refuge in the branches of a tree.

Chaucer's tale reflects humankind's need to humanise animals. The fox is portrayed as a devious crook who, while using his charm to his gain, is also prone to losing his gain to his vanity. In truth, the fox does not kill every hen in the coop through uncontrollable blood lust; instead he has found a food source that if safely cached could last many months.

The fox is reliant on his larders through the cold of winter. Items of food, such as Chauntecleer's unfortunate parents, would be buried and stashed ready for recovery when pickings become lean. The cold air and part-frozen soil slows the deterioration of cached food, ensuring that during a lengthy freeze there may still be scraps to be found.

There is no sign of a big freeze this winter. Not yet at least. And in many ways a crisp, cold but bright day would be preferable to the dull grey that we have been having. It would certainly give a useful boost to my serotonin levels. What is curious, though, is the quiet. I have reached the car and paused and there is very little going on. The woodland here is rich, with mature beech and plenty of understorey species. Fallen limbs are left where they lie and the moss on the banks looks deep enough to sleep on. Yet the air is as silent as it is still. I wouldn't expect much birdsong, although I have heard mistle and song thrushes tuning up elsewhere, but there should be a business of tits, treecreepers and goldcrests that isn't there. No rustle from a grey squirrel, no seep from bullfinch. Ah, but there is a nuthatch, somewhere in the canopy overhead. The steady plink as it slinks around the branches rather than the hurried bubble of its main call. And now the spiralling notes of a goldcrest. A little way off, too far to pinpoint, but unmistakable nevertheless.

The quiet is appropriate, though. To now. Energy is precious in winter, survival often a daily tiptoe along the edge of a knife. This is no time to be squabbling or exploring or wasting a single scrap. It might be easily overlooked from the inside of double-glazed glass and cavity wall insulation, the tweak of a thermostat more convenient than putting on a jumper. But a cold snap will still take its toll, on the homeless and elderly, those watching every single penny. These were the Dark Days, a critical time for the Celtic people. A period peppered with disease, famine and death. Fresh food was scarce, with the ground offering little in the way of sustenance. Meat could be sought, but with little light in which to venture, and harsh, cold weather, hunting carried ever greater risk. This was a time when darkness was something to be truly feared; nights were best left to the wolves and bears. A slip or trip could cost you more than a sprained ankle or fractured bone. And the quiet itself would have carried an eerie menace. Anything breaking it could carry direct threat, from something or someone desperate.

I ponder Jean-Jacques Rousseau's words taken from *The Reveries of the Solitary Walker*:

> An absolute silence leads to sadness. It offers an image of death.

I don't necessarily agree, though when Rousseau wrote these words the influence of humankind upon the environment was far less marked, not least via sound. And without the buzz of the fridge or the high-pitched whine of a phone charger, without aeroplanes and trains, and 40 million vehicles pounding on tarmac. Without televisions and radios and mp3 players – absolute silence might have impacted Rousseau differently than it does my ears.

I have only experienced it once, on a holiday of spontaneity to Cumbria with old schoolfriends Kieran and Hugh. It was mid-December and a storm had dumped several feet of snow across the fells before swirling off to the north and leaving behind high pressure and plunging temperatures. We ventured up on to Wrynose in Kieran's old Ford Fiesta, abandoning the car to the ice and reaching the top on foot. There the sun shone through frozen air and nothing moved. We lay down with our backs on the hardened snow and listened. There was silence. No sound at all. Just an eerie, beautiful quiet. Perhaps the moment itself, the company I was in, that very point of my life, softened Rousseau's sinister edge, but I was untroubled and felt no sadness.

The footprints of a fox, frozen in the snow, were the only sign of life apart from our own. The prints trailed purposefully across the snow plains, clearly made by an animal on the move. That the animal itself was likely miles away added significance to the tracks themselves. They were probably days old, made when the fall was fresh but now compacted as the snow hardened in the plunging cold.

As we stared skyward, an airliner appeared from the southwest. It was high, probably at cruising altitude, and it took some time before the rumble of the engines reached us. It had promised noise, though, an end to our silence, and perhaps that was the reassurance my mind required.

It was odd last year to see the clear morning sky devoid of vapour trails. The quiet that came as we locked down against the pandemic brought positivity as well as fear. The roads emptied, aeroplanes grounded and the world seemed to breathe once more.

'Stay at home' we were urged, and people did. It was more difficult for some than others, and we were far more fortunate than most. Our world is already quiet, tucked into a quiet corner, a walk from the car which is already parked in a cul-de-sac. And we were used to the solitude and the lack of doing. Two trips to Morrisons over the course of the week became one, and I began to forget what a petrol station looked like.

Stay at home.

Interesting. Those words resonate within other words I have been reading. My curiosity about the moon and the influence it may have had upon the Celts. *Anagantios*, a word engraved into a piece of metal nearly 2,000 years ago. I have found a variety of translations for it, several recurring, and none wholly convincing. 'To stow the harvest'; 'to escort'; 'to take a ritualistic bath'; 'to stay at home'.

A word presumably written by a people who did not write means nothing on its own. It is like trying to solve a whole crossword with only one clue. Yet it seems certain that *Anagantios* was from a language used by Gaulish Celts, found, as it was, on what is known as the Coligny calendar. A cache of bronze pieces found in 1897 in central eastern France that, when reassembled, represent part of an apparent calendar. The words and structure found upon it have since been the subject of great conjecture, with speculative meaning being shared and repeated until widely accepted. The more I read about it, the greater the confusion, but in recent discussion one name seemed to be oft-quoted.

Xavier Delamarre is a renowned French linguist and lexicographer and regarded as one of the world's leading authorities on the Gaulish language. Amid so many opinions, I felt drawn to his because he didn't really have any. Delamarre seems interested in the etymology of the words rather than their wider relevance. His literal translation of *Anagantios* is 'no go', finding roots in Welsh and Old Irish to support his theory. *No go.* Stay at home.

I felt as though the fog of information had cleared a little, but there remained a major issue. The widespread thought determined *Anagantios* to be a month corresponding to late summer, hence some of the alternative translations. Why would the Celts stay at home in August? That made little sense to me, but also contradicted my own wishes. I had seen the first full moon of the year referred to as the 'Quiet Moon' and I liked that. It felt apt. And, of course, the notion of 'no go' fits rather well with it.

I pause briefly, with the car door open. There are a few seeps and whistles from the trees, but no human sound and just the faintest rustle of air through the pine needles and stubborn cling of the beech leaves. All is a hush. The Quiet Moon indeed.

2

THE MOON OF ICE

Let us love winter, for it is the spring of genius.

(Pietro Aretino, *The Works of Aretino: Biography: De Sanctis. The Letters. The Sonnets. Appendix*)

ON A SOFT DAY, with a gentle southerly breeze, it is possible to smell the sea from our garden. It is faint, never more than a subtle tickle that vanishes as soon as you focus upon it, but it is enough to remind us that the sea is not many miles from us.

It is invariably Sue who catches the scent first. Her sense of smell is keener than mine, a sensitivity less suppressed by too many years of smoking on my part. More vitally, though, is Sue's affection for things salt-edged. Having spent her teenage years beside the sea, her dreams swirl with the tide. The sea is a friend, a confidant and an unwavering constant. After a few minutes sitting on the pebbles of Chesil, Sue will be drifting off somewhere else.

Since living within easy reach of the sea, my own feelings have deepened. I love the ever-changing mood, the flat calms of September still and the violent rage of an October storm. I love the cool edge that the air carries on the hottest of days, and the salt that I can still taste on my lips long after I've left for home.

But I am, and will always be, a mountain man. Stirred by the peaks and crags that I may not have ever lived among but that have filled my imagination since I was a small boy. I am quite particular about my mountains, too. They need to be Scottish, preferably Hebridean, and definitely have to play home to golden eagles. I can admire the beauty of any montane landscape and lose myself within it, but that is in part due to association. Those moments from my childhood that helped blur fantasy and dreams into reality were made on family holidays to the west coast of Scotland, and nothing else can ever come quite as close.

Any single missing element leaves me with a niggle of longing, similar, I imagine, to the sense Sue might feel on the shore of one of the great lakes. There would be a water-filled horizon, the cry of gulls and the lapping waves beside her feet. But it wouldn't smell right, and even though she could take much from the moment, the gaps would be filled by her subconscious. Total immersion diluted by nostalgia.

So it is that I feel slightly guilty as I walk past the bramble and gorse, down the slope, with the cold of the air freshening my face. It is early, not so early as to see the moon set, but early enough to watch the sun rise. I am not a morning person, even though I love the clean, soft feel of dawn, but Sue simply cannot make it out of the house at this sort of hour. Not unless she has a week or two to recover. But my purpose this morning is reliant upon those early rays of sunshine after a cold night. And it certainly was cold: the ice was thick on the windscreen and the frost found its way all the way here to the edge of the sea. I have not seen such a thick glisten of white quite this close to the coast before, and the temperature is still below zero. My lungs fill with the chill of the air. It feels so clean, so wholesome. That fresh edge of the sea coupled with the icy, Arctic cool that has sat rigid for several days. Bright sunshine and cold nights. It has been most welcome.

Most people struggle with the winter blues to some extent. Some might not even notice it, so shallow is the dip in mood, whereas others, myself included, can end up in a deep funk. I remember the sensation as a child, when I would completely lose my appetite. I had nothing to relate it to and no reason to try to explain it to my family or friends, but for a few days I would have to force myself to eat and even favourite foods would be a struggle. Then, when I was 10 or 11, I woke one morning and couldn't get out of bed. I probably had the physical capability, but my mental fortitude was numbed to the point where I felt paralysed. My parents despaired, and shouted, and tried to drag me out, but I simply couldn't do it and I had no idea why.

The pattern continued, always in late February or early March, and at first the doctor was unable to explain why. I wasn't aware of it approaching, but if it crept up unexpectedly, it wouldn't leave with such decision. For some days after I was out of bed and eating, I would drift around as though in a cocoon. The air would feel odd on my skin, a strange, almost clammy sense of disassociation. And I was painfully aware of it. Going back into school in such a state, with the inevitable jibes about my absence, was incredibly hard. I felt extraordinarily vulnerable.

I was nearly 17 when the doctor diagnosed seasonal affective disorder. He seemed strangely excited that he could follow the pattern back to my pre-teen years, whereas I was relieved to have some sort of validation. It made perfect sense, the cumulative impact of several months' worth of short, dark days leaving me dishevelled at a time when the sun was beginning to regain its warmth.

Although the diagnosis was something of a personal breakthrough, certainly with regard to my understanding of myself, it did not explain subsequent, darker episodes of mental health. It did, however, encourage me to try to help myself through the winter. To get outside and feel the sun on my face. Hence the fact that I will lever myself out of bed and out of the door even when I desperately don't want to.

Those 'forced' excursions are always beneficial, and almost always feel so at the time. Yet I still have to convince myself of the fact beforehand. At least, I do until the fug begins to clear. Sometime in mid-spring, near the equinox, I find myself walking without prompting. I've kicked my heels because I want to, not just because it's good for me. I might not always notice the transition, instead making a retrospective realisation a bit like when you realise that the nagging cough that rattled for weeks has finally gone.

Today, though, my walk is slightly forced. I was awake early but had just enough energy to bundle myself out of the door and into the car before I had a chance to argue. And the incentive I used to prompt myself was cold-blooded.

I don't recall it being particularly intense, but an old schoolbook of mine, from when I was 6 years old, has suggested I had something of an obsession with adders. The first page shows a drawing of an adder chasing a 'doormouse' and the theme recurs several times over, and I'm sure my fascination was due in part to fear. I remember being relieved to learn that British natural history was tame in comparison to elsewhere in the world. We lacked the giant pythons, cobras and Komodo dragons, the tarantulas and funnel-web spiders, the lions, tigers and polar bears. Our largest carnivore was the humble badger and our bats preferred munching moths to sucking blood. The only animal that posed any sort of risk to my young mind was *Vipera berus*.

My first ever sighting was fleeting. A rustle in the Purbeck scrub and the briefest glimpse of a vanishing tail. It was perfect. I had enough of a view to feel a little bolt of eye-widening adrenaline but was also reassured that my parents had been right. An adder would be more scared of me than I was of it.

I did get complacent a few years later. Running down a path between the bracken on Dartmoor and seeing an adder directly beneath my leading foot. The world slipped briefly into slow motion, and my mind whirred. I only had about 15 inches and the tiniest fraction of a second, but somehow managed to arch my back and thrust my calf forward to miss the snake by quite a distance. It was a timely reminder to watch my feet, and ever since (and fingers crossed for the future, too) I have not come knowingly close to injuring an adder.

I have developed a half-decent eye for them, though, and start my searches early in the year. Well, in truth I don't actually stop looking. Just as many anticipate the first swallow sighting of spring, I like to push back my earliest adder encounter. I have had a couple of January sightings, with the earlier on the 28th, but so mild are the winters that I doubt the adders in this part of the country manage to slip into proper hibernation. Certainly here, on the south coast, where the temperature rarely nudges anywhere close to the cold of last night.

Yet these weather conditions are actually ideal for adder spotting. The sun will shine for nearly nine hours today, just as it did yesterday. And as I sat in our lounge yesterday morning, the glass of the French doors bathing me in warmth from the low sun in the south-eastern sky, I began to think like a reptile. If I were cold-blooded, and my life dictated by the temperature of the world around me, I would have found that sunshine irresistible. Particularly after an icy-cold night which would have sent my body into torpor. The warmth of the morning sun wouldn't allow me to sink back into a state of hibernation, though, and instead I would be dragged towards it like a sailor drawn to a siren's song.

It is a process of thought familiar to my angling self. Fish too are cold-blooded and I consider what they need – essentially food and security in efficiency. And so I learn to read the water, to work out where the fish are most likely to be lying. Looking for signs of weedbeds and undercut banks, creases on the surface where fast water meets slow. When a life is dictated by such subtleties of the environment in which they live, it becomes slightly easier to predict behaviour.

So it is that I don't necessarily 'think' as a fish would, but I consider how the variables of air and water temperature, light levels and wind direction might drive their response. I apply a very similar approach to reptiles, and by 'thinking' like an adder, I can often find myself one – especially in the early part of the year. Adders will slip out of their winter holes on a mild, sunny day, but always keep within a tail-swipe of safety. Find one in late winter and there it will stay until the urge to mate prompts it to leave its hibernaculum and venture forth at the behest of its hormones.

I am confident of finding an adder this morning. The frost is quite thick in places but is ceding as soon as the sun touches it. The birds are certainly stirred; a stonechat is clinking from a hawthorn on my left and a couple of linnets zing above my head. It might be they who set off the skylark, who rises from the pasture only 10 yards or so away, breaking into immediate song. What a sound. The way it holds on to the higher notes and seems to drop wheezes and whistles beneath. Almost reminiscent of the breath of bagpipes or an accordion, that constant movement of air creating multiple layers. One skylark soon becomes three, with another ascending over the longer grass further to my right.

I don't recall closing my eyes, but the clink of a gate has broken a spell that I had drifted under. I had floated off with the larks, felt the sun on my wings as I lifted into that beautifully pure air. I was more likely to break into snore than song, however; my subconscious is grumbling at me for keeping it awake for too much of the night and it spied an opportunity to make up for lost time.

Perhaps I do myself a disservice. Sue taught herself mindfulness as a means of coping with chronic fatigue, and she has often noted how I seem to find myself in that state of mind without any effort. It is unwitting on my part, although something I have always done. The standard childhood punishment of being sent to sit on our beds was glorious in my mind. I could kick back, fondle the bumps in my sheets and stare blankly out of the window. It was pretty much my perfect way to spend an hour. In recent years it has become rather tainted by technology though. Rather than step sideways into a moment I reach for my mobile phone. Then to check a cricket score in a match of no relevance or scroll through a dozen social media posts without reading a single word. Staring at a small screen probably contributes to my erratic sleep patterns, although last night's fits and starts seemed undoubtably linked to the moon. There is a chance that it was psychosomatic; after all, the moon rose before dark last evening but then dominated the empty sky. I was well aware of the light when I woke through the night, but it was the strange energy that impacted more. I have taken a mild sedative to help me sleep since I was a teenager, and my mind will buzz without it. But the sense I had last night was a step beyond that. I felt wired and alert, in a similar way to the wide-eyed pulses that thump during an electrical storm.

The moon has long been associated with erratic behaviour, of course. Peaks in violent behaviour, prison unrest and even birthing have long been associated with a full moon. The word 'lunatic', its etymology fairly self-evident, has been not just a part of our language for centuries, but also used within the structure of mental health. In fact, it was only under Barack Obama's administration in 2012 that the House of Representatives removed the term from United States federal law.

There is no scientific evidence behind a correlation, and studies seem to throw up very different results, but such strength of evidence in anecdote cannot be disregarded. I might, perhaps, be willing to overlook the tales of lycanthropy, although having watched *An American Werewolf in London* I will definitely keep off the moors should I go wandering under a full moon. Unlikely too is the oft-pitched theory that the human body, being 80 per cent water, is manipulated by the moon in the same way as the tides. This notion is often attributed to Pliny the Elder, and his epic tome *Naturalis Historia* is littered with references to the moon and its importance. So many aspects of health and day-to-day life were dictated by the position of the moon in the sky. Ailments could be cured with ghastly sounding concoctions, often involving the brains of animals, but only if they were gathered or administered on specific points of the lunar cycle. Apples should only be picked after the autumn equinox and between the sixteenth and twenty-eighth days of the moon, whereas timber should only be cut after the twentieth day. I cannot vouch for Marcus Varro's assertion that cutting one's hair immediately after a full moon would caution against baldness, and I fear it is a little bit too late for me to find out. Pliny wrote also of the Egyptians' obsession with the lunar cycle, particularly when forecasting weather, and of how the growth patterns of shellfish would mirror the form of the moon.

For balance, it should perhaps be noted that Pliny also observed that the menstruation of a woman could cause crops to fail, wine to sour, weapons to blunt and even swarms of bees to instantly die, so perhaps we should treat many of his records as representative of a time. What is in no doubt, however, is the importance of the moon across many different cultures and belief systems, and one particular comment caught my attention as to explain perhaps why. There is no doubt that modern cultures, living within such extensive light pollution, have a lack of awareness of the night sky. Whenever we are visited by friends from London, or anywhere where the streets may be artificially lit, they are invariably struck by the clarity of the darkness. Not just the moon, but the complete sweep. The thick soup of the Milky Way and the hypnotic scintillation of Sirius. If Jupiter is prominent then minds are blown if I pass my binoculars and point out the moons (normally four) that are visible around it. For the most part, though, we live in light brighter than the stars can ever offer and draw the curtains as soon as the sun sets. We are not so much ignorant as oblivious. But one thing that we take for granted, particularly in the Western world, is water. Turn on a tap, pull a lever to flush a toilet, turn a dial to make the shower slightly warmer. A drought might yellow the lawn or curl leaves in the vegetable patch, but we are not really inconvenienced.

The comment from Pliny might not have been meant to carry any great weight, yet it led me into thought:

> the sun absorbing water while the moon gives it birth ...

It seems rather obvious, but at a time when every drop of water was so precious, the motive of the sun, driving away the clouds to evaporate and desiccate, would seem rather sinister. It might provide light and warmth, but it also takes away something that without which we would quickly perish. Yet on those scorching days of high summer, cloudless and blue, the moon does what it can to compensate the aridity. As the sun cedes the sky to the night so dew falls from nowhere. A damp blanket gifted from an unseen source. And the fact that the moon is responsible for the rise and fall of the sea points towards a sentient spirit with goodly intent. Giving birth to water and in turn to life.

The heat of summer feels a long way away today, although it is lovely to feel the warmth of the sun. The frost is holding firm in the shadows but is glistening back to dew wherever the sun can reach it. The walk to the back of the beach is only short, a matter of minutes if I don't dawdle, yet the day has already gained momentum. The air is full of sound. Chaffinch, stonechat, linnet, wren, robin – all singing beneath the unbroken wall of skylark. A song thrush is distant, but a blackbird sharpens within comfortable earshot, and I have to give him some time. His song, the deep and deliberate fluty notes, always captures me in the early part of the year. He sings sooner in towns, influenced perhaps by the artificial light and that additional warmth, but I am not used to his song until winter is easing. This bird is still a little rusty. His voice needs some tuning and he does not sing with the bold confidence that will come. But those slightly lazy deeper tones, interspersed with a softer, sharper whistle and flick, are worth a listen even if it is only a soundcheck. 'Blackbird' might seem a perfunctory name for such a singer, but there is no arguing with the description even if it doesn't fit with the dark brown of the females and juveniles. The orange of the beak and the ring around the eye are tangerine-sharp in the male wearing his spring moult, though, and the gloss of the plumage is rather splendid, too. It is not a term I have come across since we moved this way, but a traditional West Country name for the blackbird is a 'colly' or 'colley', derived from the coal, soot-like colouring. And aside from a couple of dozen making it into a pie, the colly bird also made it into a well-known Christmas song. The 'four colly birds' in 'The Twelve Days of Christmas' more recently evolving into 'calling birds' – the latter being the lyric that I have always known. The progression of language so often goes unnoticed. A mispronunciation that slowly manifests like those false facts or moon monikers.

I have reached the back of the beach and it is at this point that Sue and I usually part – she continues down on to the pebbles to sit and watch the waves beyond, whereas I head sideways to explore the edgelands.

As sea levels rose at the end of the last Ice Age, sandy deposits in Lyme Bay were steadily eroded and the resulting particles were rolled inshore to form a barrier beach. The stones were shaped and smoothed as they moved against each other, forming pebbles that the tides piled high and straight. Chesil Beach has changed little in the last 5,000 years, although it seems to be creeping slowly eastwards. For around half of its 18-mile length, the beach protects a vast saline lagoon, the Fleet, from the pummel of the sea. A narrow channel at its eastern end links the Fleet with Portland Harbour, ensuring a tidal influence and allowing a movement of fish. The lagoon is shallow, rarely deeper than I am tall, and its serene and slightly benevolent atmosphere contrasts with the wild of the Chesil's seaward side.

The tides here are strong, and the undertow, particularly with the added weight of a spring tide, potentially lethal. Shipwrecks litter the inshore and bathers and divers are too often caught unawares. Despite this, the beach is rarely empty. Anglers come throughout the year and, despite the lack of sand, tourists flock in their thousands – particularly during the summer months.

I occasionally fish, mainly for the mackerel that chase the whitebait shoals right into the breaking waves, but I more often visit to walk behind the beach and among the links.

I am towards the western end of Chesil, a few miles from the Fleet. The meadowland is occasionally grazed by sheep and cattle, but is, in the main, uncultivated and unimproved. Here, the salt-edged wind checks the arboreal growth, leaving hawthorn and blackthorn hunched and tight, keeping their heads down and tangling themselves into a thorn-laden mesh. This creates a narrow strip of wild that links the beach to the land and resists the influence of both. Too salt-stained and storm-beaten to be tamed for agriculture, but earthen and solid enough to hold firm against the waves. A glorious no-man's-land where almost anything seems possible. Sometimes only inches wide, a narrow clasp of sea beet and samphire, there are points where it stretches itself further inland, creating pockets of habitat with a ribboned link between. As spring progresses, this is where scores of migrant birds touch down, finding immediate safety in the tangle of twig and thorn that offers sanctuary and food for those arriving from Africa. Whitethroat and lesser whitethroat, whinchat and wheatears. Chiffchaffs, willow warblers and blackcaps. The links come alive and offer similar protection in the autumn for those same birds and their offspring readying to leave.

For now, though, the resident birds have the links to themselves. Dunnocks, robins, wrens, song thrushes and blackbirds. All in voice this morning in varying levels of commitment. There are tits too, blue and great, and the stonechats, linnets and yellowhammer. Quite a list when I begin to mentally tot them up. There are a couple of species that I hear rather than see. I did catch a glimpse of a Cetti's warbler on my way down the slope, but several more have shouted their song from cover, one right next to me, without showing so much as a feather. Unlike the warbler species that will soon be winging in from the south, the Cetti's remains here all year. In Britain, it has benefitted from climate change, and expanded in range and number. I remember reading about them as a child and feeling they were a bird out of my reach, even though their stronghold was in the south. Today, though, they are a familiar part of the soundscape, especially near water. The song is loud and incisive. A first sharp note followed by a knife-carving flurry, a sound that demands attention. I have often been asked (a pair of binoculars around the neck always gives the impression that I might know what I'm talking about) by people as to the source of a Cetti's song. Walkers who might normally stroll in relative oblivion enjoying the birdsong but struck by the voice that seems intent on drowning out the rest.

The stretch of the South West Coast Path that I am briefly joining is well trodden, and although I am often asked to identify a Cetti's, more often people are curious as to what might be grabbing my interest in the thick of the scrub beside my feet. 'Adders,' I smile, often drawing a reaction of slight shock. I like to highlight these snakes though and point them out whenever I can. The fact that people can walk past an animal and be completely unaware of it helps to demonstrate just how timid these snakes are. And few creatures in Britain elicit the kind of response that an adder does. A recoil seems more likely to stem from fear rather than disgust and is often accompanied by a sense of awe (such as from my 6-year-old self) and curiosity. It seems that the majority of people have not actually encountered an adder, and they are a species that we are increasingly unlikely to meet in serendipity.

The adder is traditionally an animal of broken woodland and scrub – habitats that are increasingly tidied away to nothing. And they are not pioneers, taking their chances in crossing inhospitable land in the hope that there is somewhere suitable beyond. Instead, they find themselves marooned in small, shrinking pockets. Isolated populations without fresh genes to splice but a relatively long life (an adder may live beyond twenty years) giving a false impression of permanence until it is too late. Gradual declines tend to go unnoticed, particularly when comes the annual August melodrama of adders biting anything that moves. They will bite, but only ever as a last resort: venom is too precious to waste on something that isn't going to then be eaten. But dogs will always investigate with their noses and holidaymakers walk in sandals or flip-flops, providing editors with something sensational to fill the columns during parliamentary recess. Too many people hear of only this, and the negative image of an evil snake is further embedded. They have always been a timid creature, though, as noted by Victorian naturalist and writer John George Wood in his 1879 book *Lane and Field*:

> I passed some time in Hampshire in a spot celebrated for Vipers.
>
> At first I was rather nervous about them, as it was hardly possible to take half a dozen steps without seeing one or two Vipers gliding away. But they were so shy, that they retreated at the sound of the footsteps, and in a day or two I thought no more about the Vipers than about so many earthworms. Indeed, they were much more afraid of me than I of them.

Such an abundance of adders seems inconceivable today, here or anywhere. I have just walked a 100-yard strip of bramble and grass, where I have found up to four snakes present in previous years, and seen not a scale. It is surprising. A couple of spots have grown out, stalks stretched, meaning there is insufficient cover tight to the ground. Spots where adders would, in the recent past, have spent the winter now not offering the same level of safety. But there is still plenty of scope and enticing hibernacula – just no snakes. It remains cold, the air temperature is probably only just above freezing, but the sun is warm and as I touch the grass on the raised bank it feels quite cosy.

I've turned my back to the sun and am leaning against the gate which breaks this run of adder scrub. A pair of stonechats are eyeing me warily, both perched on the spindled remains of what may have been gorse. They sit in typical stance, one leg bent and raised, the other straight, just as I have unwittingly done – my right foot rests on the lowest rung of the five bars as I hunch over the top. The male bird flicks his tail several times, but within a few moments both seem to have settled and decided I am not an immediate threat. They look glorious in the low-level sunlight, the rust-orange on the female's breast almost as rich as the male's. She lacks the black head and smart white colour of the male, but instead her eye and the flush of her cheek are more prominent.

As I watch them, another colly bird pipes up from the thicketed hedgerow to my right, slightly sharper than the first but not quite liquid blackbird. His is not the only name to have changed over time. The snake I seek was originally called *naedre* in Old English, a word that became 'nadder' (and shared a name with a Wiltshire chalk stream) and then lost the opening letter to the preceding article to become 'an adder'. The same process led to an 'ewt' becoming a 'newt'. The evolution of language is as fascinating as it is relentless but can be met with nostalgic reluctance, perhaps reflected in generational resentment. Change can be difficult to adapt to, so we tend to blame those who we perceive as forcing it upon us. And we all have our own compass, similar to that of morality, which we accept. There was terrible music in whatever era defined us, but we only remember the tunes that soundtracked moments we treasure. And as much as I resist text-speak, and spell out every word, I do not insist upon placing an 'e' on the end of 'old', writing 'ye' instead of 'the' or insisting that a coil of serpent scales be called a 'nadder'.

As long as the meaning is clear, then all but the most ardent pedants can relax, although I am finding frustrations in the lack of clarity within the Celtic language. Take the Durotriges, for example, the name given to the Celtic people who inhabited Dorset. In his 1607 work *Britannia*, William Camden suggested it meant 'dwellers by the water or sea-side' – having been born from *dovr* or *dwr*, still used in Welsh for 'water', and *trig* meaning inhabitant. Alternatively, the term 'duro' might have come from the Latin for 'stiff or hard', pointing to 'dwellers of the hard ground' or 'fort dwellers'. Any of these translations make sense and perhaps only the origin of the word itself would give the definitive reasoning. Did the Durotriges call themselves as such, or was it a name given by the Romans (and therefore reinforcing the likelihood of the Latin meaning)? And, again, how much importance should we place upon it?

As I have delved deeper into the etymological mechanics of the Coligny calendar, another thought struck me. Were this civilisation to end today, and in 2,000 years the only surviving calendar was the one hanging on our kitchen wall, what would the finder deduce? Aside from the presumption that freshwater fish (that being the theme of our calendar this year) were held in some esteemed regard, they might piece together the language found and perceive it incorrectly. September and October relate to the seventh and eighth months, December the tenth. With no reference to the Caesarean influence upon the naming of July, and subsequently August, the Julian calendar would quite reasonably be deduced to have begun on 1 March.

An idiosyncrasy widely accepted of the Coligny calendar is that it was based upon lunar, not solar, cycles. Therefore the period of a year that we know today – the time between vernal equinoxes or, more familiarly, the time it takes the earth to complete a full orbit of the sun – cannot fit into the structure. The moon completes its own cycle in around twenty-nine and a half days, meaning twelve cycles (equivalent to the twelve months of the Julian calendar) would leave a 'year' around ten days short. Instead, it appears that the Celts followed a calendar that ran over the period equal to five lunar years, with 'leap months' making up the gaps similarly to the 'leap day' added by Pope Gregory XIII in the sixteenth century to the existing Julian calendar.

The Celtic calendar would still need a starting point, of course, and that is something that has become increasingly contentious. The month of *Samonios* is the accepted beginning, but its physical position is disputed. A connection to summer, from 'samo' and various derivatives of it, was quickly established, and the apparent importance of days within *Samonios* suggested a major festival or event. Popular thought pointed to the summer solstice, an event deeply associated with pagan and druidic culture. Even today, tens of thousands of people gather at Stonehenge to watch the sunrise on the longest day of the year, a moment that many believe was integral to the positioning of the stones themselves. The importance of the solstice seems to have been such an accepted notion that much of what has been written and researched around the Coligny calendar begins from that viewpoint. Translations and configurations are worked back to the belief that *Samonios* corresponds to the Julian month of June. Yet such thinking would mean that we 'stay at home', beneath the Quiet Moon, in September – a month followed by *Ogronnios*, which is almost invariably etymologically linked to 'cold' and 'ice'. Surely a month of winter rather than mid-autumn?

So, what if *Samonios* falls later in our solar year? What if the festival it relates to is Samhain, which was celebrated around the time that October turns over to November? A celebration that is documented in Gaelic culture from the ninth century, suggesting a deeper and longer association within the unwritten world of the Celts and a very obvious connection with the root of *Samonios*. This line of thought has become increasingly visited, and much of the more recent thinking follows this link. The Coligny calendar would have been produced at a time when the Celtic influence within Gaul was waning. The language was far more likely to have been bastardised and conflated with Latin and Greek. But the core elements of the Celtic tongue would have survived longest within the languages of the retreat. Of Ireland, Scotland, Wales and

Cornwall. Those languages will have evolved as all do, but the festival of Samhain can surely be linked to that carving in bronze, no matter what linguistic influence was laid upon the person doing the carving. The Coligny calendar was a functioning piece of apparatus, not just a piece of decoration, so it has vital value as a source even if it is unlikely to ever be fully understood.

Such thoughts float around my head on a cold, icy morning in late winter. I am leaning on another gate. This time without an entertainment of stonechats but with an increasing mixture of surprise and concern. I have just walked my banker stretch: an area set back from the path, where there is a shallow plateau of grass that is back-edged by bramble and thorn. The step up is just high enough to deter any free-ranging dogs and, as a result, this 50-yard run is almost completely undisturbed. In past winters, this has been the most reliable spot for adders that I know locally. A couple of years ago there seemed to be a hibernaculum shared by several mature females. I found three within inches of one another and a different snake a few yards away. In another ten paces lay a young male, almost certainly unaware of his potential fortune. Female adders do not breed every year, but one of those four would surely have been willing to mate, and providing the male didn't dawdle he would have an instant opportunity to offer his credibility as a suitor. Male adders are noted for their early-spring 'dance', a behaviour attributed to competing individuals likely driven by rising levels of testosterone. Sadly, it is something I have not seen, and the opportunity will only lessen as the adder drifts further into scarcity. I did hear of a woman witnessing a dance a couple of miles along the coast just a week or so ago, so there is a chance, but nothing compared to what might once have been. Another woman, who I met elsewhere while addering, told me of her experiences as a child in this same spot where I am today exploring. She recalled 'great balls' of writhing males – several dozen individuals – and, even allowing for the distortion of memory over time (I didn't ask, but suspected her to be nearer 90 than 80), she was witness to a wildlife spectacle of old. Of the past.

Having blanked at my banker, I banked left and followed the curve of the links slightly inland, where I have not searched before. It has all looked promising, but this gate is propping up more than just my body. I feel weighed down by that mix of emotions. It was nice to let my mind drift off with the Durotriges, but now I am back in the present I feel encumbered by the reality of silent disappearance. The sunshine and exercise will do me good, as have the larks, chats and collies. But I always get something extra from spotting my first zigzag of the year. It is a different kind of encounter with an animal driven by different needs. Something happens deeper inside me, something that will float me through the rest of the day and into the next.

There is a line of gorse further up the slope, though, cutting a rough path along a shallow depression in the ground, which seems to look more promising the more I look at it. There are patches of thick grass and tangle which the sun seems to be gently nudging into warmth. This is the spot. Surely. I scale the gate and work my way to the top of the gorse run.

My shadow is something of an issue, and something I should have considered. Had I headed to the bottom end and worked up the slope then I would have been stepping ahead of it, but instead, it is falling exactly where I don't want it to. I edge backwards and line up my binoculars. I could, of course, do the sensible thing and start the process again from the more logical beginning, but that would mean doubling back on myself, and regardless of what I see in the next ten minutes, I do feel I need to be cutting a path back to the car.

I don't like scanning for adders through binoculars, although I might if the image were a little crisper. As it is, the view flattens and the little bumps and openings disappear. More often than searching for a snake itself, I look for a place where I would expect to see one, that spot a flick away from sanctuary, south facing, and sheltered from the breeze. At this time of year adders can be reasonably predictable, another reason why I have been concerned by their scarcity. There though, a little further on, is the perfect spot. A slight cove in the line of gorse that was likely once the run of a fox or badger. Now regrown just enough to leave a slight step a foot or so above the pasture, and thick with cover. Despite the lack of success up to this point of the morning, it is not a surprise to focus the binoculars on a snake. A male. Not a large adder, but they are not a large snake, and for once, I am glad for the added magnification. I can study this creature without any risk of disturbing it. They can be incredibly tolerant and have no wish to move and waste vital energy unless necessary. I could approach this snake to within touching distance and it would likely stay put. Provided my movements are slow and I breathe only through my nose. The tongue, slowly flicking, is tasting the air and the waft of adrenaline unwittingly released by a potential predator would cause flight. A more frequent interaction would be with deer, rabbit, sheep or cow. Animals too big to eat and, aside from the squash of a heavy hoof, no threat to a snake. Just as I think like an adder, I also act like a benign, cud-chewing cow with no care for reptiles. For now, though, I will remain slightly distant and absorb the moment. The scales of this adder are well dusted from a winter in subterranean sanctuary, and the indent surrounding the eye is fairly caked. But gosh – that eye. Deep red, flaming around a narrow vertical slit. Unblinking, Tolkien's all-seeing, it is the fire to the ice of the morning.

My heart leaps.

3

THE MOON OF WINDS

Storm-clouds whirl and storm-clouds scurry;
From behind them pale moonlight
Lickers where the snowflakes hurry.

(Alexander Pushkin, 'Devils')

WE HAVE A CURIOUS relationship with routine. It is something by which we thrive, a security of structure in which our basic needs are met from birth until death. We eat, drink and sleep around school timetables and a nine-to-five, with our evening's entertainment packaged neatly into half-hour segments, the lives of fictional characters seeming more valuable than our own. We have a favourite mug or the usual tipple, a takeaway on a Friday and a roast dinner on a Sunday. We wheel out the bin the evening before a collection and see the same faces in the supermarket as we push around our trollies at the same hour each week. We park the car in the same spot and use the same pump in the same filling station. With everything in its right place, there are no alarms and no surprises, and that suits us. That allows us to cope.

Stepping out of that routine then becomes a luxury. A lie-in on a Saturday morning or a late-night film on Saturday night. A midweek drink in the pub or a meal out. We long for our holidays when we spoil ourselves and revel in doing either nothing or everything. Stepping back into normality comes with a slap around the face, yet it takes very little time to settle back and feel the slightly deadening reassurance of familiarity. We trundle gently through the weeks like a rickety train on an old, forgotten track. Stopping at every station, just as it always has, but no longer aware if any passengers are getting on or off.

It is odd that we find such ease in switching off but boarding that rickety train provides us with mundane comfort. By positioning our needs within easy reach, we don't have to worry or find ourselves in situations where we are forced to think. It is almost as though we become institutionalised within time. We count down the days and hours until those points when we allow ourselves to step sideways and break the mould, and in the meantime happily allow our lives to be dictated by society's need of us. Such a process works, for many of us at least, but the payoff can be harsh. As we drift in conformance, so we lose our ability to be. We close off the parts of our mind that question and engage, in order to make the passage less fretful, but too often we forget to open back up again. Then, when presented with an opportunity to stop and absorb, we have completely forgotten how. And the loosening of the restraints leaves us unsettled and slightly vulnerable, so we hurry back to that rickety train and lull ourselves to that familiar rhythm. Not that such a journey is without benefit, though. The mental strain of lockdown can be compounded by a lack of physical exertion. Working from home, or not being able to work at all, eliminates the forced commute from which we gain exercise and exertion, even if we resent it at the time. Going out in the cold or rain, when the wind is blowing hard or the sky is grey, can be too easily avoided when you don't have to do it. But sometimes pushing yourself out into inclemency can provide unexpected pleasure, especially when your normal patterns have been disrupted.

I have done just that this afternoon. I have paused for a moment, part leaning on a gate but as a means of support rather than a prop to sun-soaked idleness. I have my phone tucked into the dry of a pocket and a microphone lead to record a few thoughts. I have my binoculars too, but I hung them around my neck before I put my waterproof coat on. It is blustery rather than stormy, the rain fairly light but peppering quite hard with each gust. And I'm having to turn my back on the most desirable view, back across the hollow bowl towards our cottage, because the weather is channelling up the combe from the south-west and slapping me in the face. The main slog of this walk, along part of the Roman road that once linked Dorchester and Eggardon Hill, will be face-first into the worst of the weather, but I have already completed the most taxing part. The slope up from the village is steep and I pressed quite hard. In part to make sure I didn't give myself the opportunity to turn around and head back to the warmth of home, but also to wake my body up and remind my heart to thump. I am feeling heavy and slow. The gatepost is a welcome crutch as I slow my breaths and calm my body.

I am rarely in good shape, and have carried a belly since my late 20s, but my routine normally keeps me within half-decent condition. But I have been sluggish since I searched for adders, two weeks of minimal effort at a time of the year when I am normally thirsting for time. As winter edges towards spring there is so much to see and so many omens of what is to come. The first celandine, the green shoots of blackthorn. Bumblebees dozy and brimstone butterflies sharp in the sunshine. With the traditional coarse fishing season ending in mid-March (I still recognise the close season, despite its abolition on still waters), I have too much to do, and that works very much in my favour. There is so much going on that I can generally keep just busy enough to keep out of the shadows and myself afloat. And the places where I most enjoy fishing, the quiet backwaters and untrodden banks, require the joy and effort of finding. Rivers offer much to the angler in winter, in terms of not just opportunity but the delivering of energy. A pond or lake that steamed in mid-summer now sits sullen and grey. A surface of silent glass, quiet and eerie. When the temperature falls, the water disappears altogether beneath an impenetrable frozen shield, yet a river flows unerringly. There is a vibrance to running water that remains whatever the day may bring and seems to pull life to it. Even on a cold day there might be hatches of flies, and the flash of a kingfisher is never as sharp as when pitched against the colourless dull of a dank January morning. The angling is active, too. Running a float along a glide, constant tweaks and adjustments, reeling in and casting. The perfect way to stay warm and engage.

Unfortunately, the rivers I fish have been out of reach since the turn of the year. I don't travel vast distances to fish, but my favoured spots are over 20 miles away – too far to be considered 'local' in the context of lockdown. And, in truth, it hasn't been a hardship. Much as I am fortunate with the world outside these windows, I can go and fish a run of ponds that are just a couple of valleys away. The fishing has been good too, with a run of specimen perch (a favourite winter species) that I would have been unlikely to catch elsewhere. Yet I have missed the benefit of being beside running water, and with just a few days of the fishing season remaining, I am aware that the slight change to my winter routine has left me mentally vulnerable. It is an old angling cliché, but there really is more to fishing than catching fish.

My walk today is, therefore, somewhat pre-emptive, and the very fact that I am able to be proactive suggests that I will fumble my way through the last linger of the winter. It does mean that my monthly moon walk has fallen out of step. The full moon, which has prompted my previous couple of strolls, is still more than half a cycle away – in fact, the Moon of Ice is still on the wane. But that could be rather appropriate.

My interest in the relationship between the Celts and the moon was somewhat whimsical. As I tread the local land so I think of those people who once shaped it and how they would have lived. And the mild frustration of the modern ignorance of those people, through no fault other than the lack of written record, became reflected in an annoyance at the adoption of ancient cultures other than our own. It is not an irritation born out of patriotism, although the subsequent dilution of fact is certainly frustrating, but more a slight embarrassment at the appropriation of other societal traits or values as a source of amusement or self-identity that might mock that original source in the process. This is not necessarily a reflection of influence in architecture, art or literature, but an ignorant glorification of lifestyles that are often minority-based and in reality a lot less glamorous than the portrayal suggests. The small-town middle-England youth purporting to an inner-city gang, or the romantic presentation of empirical dominance or oppression. The fashionable Victorian obsession with Native American motifs and beliefs seems somewhat disrespectful considering the systematic persecution of those people by European settlers. If we do not feel the shame of cultural appropriation, then we should at least recognise the history beneath.

There remains plenty of value within modern reinvention, though. In the eighteenth century, for example, the British fascination with landscape gardens may seem indulgent, but has also created monuments to monuments. Buildings inspired by the Italian Renaissance or Roman architecture still stand as easily accessible examples of a history that was not our own. We might learn and appreciate, by default, people inspired by other people once inspired.

While here in Dorset, where it almost comes as a mild surprise to clamber up a hill and not find Neolithic or Bronze Age earthworkings at the top, are places that we may not fully understand but are still able to feel. The Roman road, on which I will shortly walk, may not provoke much thought to a driver, but with feet following the paths of those who originally built it, there seems a sense of connection. At the very least, it shows that the hill fort at Eggardon was of sufficient value to have a road constructed to it. And much as I might bemoan the fake news written of the Celts by the Romans, there is undoubted value to be found.

Pliny the Elder, for example, offers a clue as to the machinations of the Coligny calendar while writing about mistletoe in chapter 16 of *Natural History* (translated by Bostock):

> The mistletoe, however, is but rarely found upon the robur; and when found, is gathered with rites replete with religious awe. This is done more particularly on the fifth day of the moon, the day which is the beginning of their months and years, as also of their ages, which, with them, are but thirty years. This day they select because the moon, though not yet in the middle of her course, has already considerable power and influence; and they call her by a name which signifies, in their language, the all-healing.

The 'robur' is an oak tree, while 'they' were the druids, from the Latin *druidēs*, the name given to the high-ranking members of Celtic society who Pliny described as 'magicians'. The influence of the druids waned in Gaul beneath Roman occupation, but the referral to the 'fifth day of the moon' and the importance of it is an interesting point. That the Celts began their months and years on the fifth day of the new moon suggests that the full moon was not the point of the lunar cycle upon which they placed most credence. It seems likely that the relevance of the full moon has gained greater meaning in more modern society and for no other reason than it is more noticeable in the sky. Although the links to lunacy and sleeplessness should not be discounted, in the twenty-first century, in the West at least, we only really notice something natural when it punches us on the nose. Those big moon-rises of autumn cause a day or two's interest, and a raft of social media posts, before we forget all about it.

So my previous walks, which I tried to link to the full moon, were baseless in that aspect. And today, with a walk motivated by my mental health, is inadvertently appropriate. A new Celtic month would have just begun.

Before I leave Pliny and wander off into the squall, I must mention a reference made later in his mistletoe writings. Pliny observes that 'Clad in a white robe the priest ascends the tree, and cuts the mistletoe with a golden sickle', something that resonated with my 10-year-old self. I devoured the *Asterix* books as a child, and still enjoy them now, but had no idea of the depth of historical accuracy that the authors, René Goscinny and Albert Uderzo, had reached. In the second book, titled *Asterix and the Golden Sickle*, the village druid Getafix breaks his sickle and is unable to cut the mistletoe with which he makes the magic potion that gives the Gauls superhuman strength. I learned a great deal of history from *Asterix*, much of it unwitting until we studied the Romans at secondary school and I wowed my teacher with a knowledge inspired by a cartoon book – the fall of Vercingetorix after the siege of Alesia gaining particular plaudits.

Another point of interest that might be prompted by *Asterix* is that of druidical influence. As the Roman influence became more ingrained within Gaul and England, and the druids retreated into the north and west of the British Isles, so their legacy was left to document by the likes of Pliny. That Goscinny created Getafix in such a similar manner to the sickle-wielding magicians that Pliny described is no coincidence. Equally, the passage above, being made in relation to mistletoe rather than a direct critique of the druids themselves, might be taken more literally as a result. Records made in part as incidental notes perhaps carry more weight than a specific record created with subjective purpose. Yet within Britain, away from those areas subject to Roman occupation, the druids remained revered until the arrival of Christianity, whereafter the Celtic influence endured. Although likely to have been indulged to some degree by the same romantic reflection that saw Native American motifs adopted, Celtic Christianity developed a style of symbolism very different to Roman Catholicism. In turn, existing beliefs and motifs were likely to be adapted rather than shunned – the festival of Samhain being an example. It seems perfectly reasonable, therefore, to consider the thinking of contemporary druids and spiritual thinkers. Such minds are likely to lean towards the support of personal opinions but will be no more subjective than the Roman and Greek scribes to whom we have paid so much heed. Anyone who immerses themselves in a subject or belief will have depth of knowledge, even if others are suspicious of the way it is presented.

I am, therefore, while in the process of disregarding various mines of disinformation, also following any interesting stream of thought back to its source. In large part because, as I have tried to find a definitive voice, so I have realised there probably isn't one, and as a result, I am beginning to fill in the gaps for myself. Often unconsciously, but also undeniably. It is a curious paradox of my own making and I have no idea what it will eventually mean to me. In the short term, though, I am finding happy distraction, and that is stopping my mind from slipping.

The rain is pulsing with the wind and, while it remains fairly light, the power with which it thuds into my coat is contradicting the material's waterproofed claim. There is surprisingly little water on the road, no great puddles or muddied sloshes, perhaps because the wind is picking it up after it has tried to settle. If a raindrop cannot find a footing as it lands then it is lifted and slammed into something else. Mainly, it would seem, on to the lenses of my glasses. I fumble for a dry piece of T-shirt to try to clean them off but the moisture from my hands merely smears a different problem. I am not sodden, but everything is damp and my main concern is keeping my phone and binoculars as dry as I might.

I could take my glasses off – I can see well enough without to walk quite safely – but I daren't miss anything. There are a few gaps between blotches through which I could make out the hunched form of a hare or a distant blob on a fence post. I am not expecting much today, but a good-sized flock of buntings has been wintering in one of the fields ahead and they might be obliging given the weather. Staying put rather than taking flight into the bluster.

Oh. I've been stopped in my tracks to look at what once was and wind myself up. As the road dips and lifts, the crease to the right runs down to a plantation of conifers, and until this winter there was a lovely area of thick scrub of perhaps an acre. It was mainly bramble with a few thorns and a run of hazel, and was absolutely impenetrable. By now it would already be sung from by blackbird and song thrush, and in another month they would be joined by blackcap, linnet and whitethroat. I would pause in this spot to listen to the jumble of song and smile. A tiny little pocket of noise, full of food and perfect sanctuary. I have no idea how many nests were tucked away in the twist and thicket, but this year there will be none. All gone for the sake of a few extra bushels.

I press on up the next slope and try to spin some positivity – there is plenty to be found. The local farmland is relatively rich in wildlife, certainly when compared to the great arable fields that I left behind in mid-Hampshire. The landscape helps, the undulations and steep slopes, boggy mire and rocky outcrop. This area is a geological melting pot, and the coastline world famous for it. It must make it tricky to cultivate though. Aside from the physical difficulty, a field thick with flint nestled between loose greensand

and the dense clag of clay, are the acid levels within the soil. A local down, noted for its flora, is peculiar in that it is a huge chalk mass topped with a thin layer of greensand – a sweet vanilla sponge cake iced with anchovies that leaves acid-loving plants growing beside grasses of chalk downland. A sense of contrast that echoes the wider landscape.

It is easy to bemoan the removal of scrub (as I have just done) or the flail of hedges, but some level of management is essential. Rather like the gorse that grows too leggy for adders to hide beneath, the species that make up a hedgerow will all reach up and out, in a process of natural succession, if allowed. A line of trees offers plenty, but not the dense cover and protection that so many species depend upon, and is of no use as the livestock-containing barrier for which it was planted. Cutting hedges back can ensure thicker growth and considerably prolong the life of the plants themselves. Providing some thought is allied to the process, that sides are cut alternately and that the depth of that cut is varied. Flailing to the same point every year will create knots and cause stress. The effort and energy required to regenerate, particularly when the stems are split and splintered, do the hedgerow no long-term good. It needs to be allowed to breathe – to reach and flourish for a time unhindered. Unfortunately, we too often take a singular approach of cutting back to nothing and individual stems begin to wither and thin. This is invariably a result of economic pressure rather than ignorance, a pattern repeated throughout agriculture. Pressures are huge, with increase in demand not always met by a fair price given. We want our farmers to use fewer pesticides and fungicides, to leave wider field margins and thicker hedgerows, but rewards in the way of subsidies might not equate to a reasonable alternative income and the methods of determination are vague or open to exploitation. Decisions are made for ease of process rather than with any consideration of specifics; a set of figures that flash up on a computer screen in Whitehall do not represent the complexities of managing a Cumbrian hill farm. Always money talks. Politicians bluster like the wet of the wind today, but thinking is always short term and too often to the benefit of the wrong people.

Hmmm. So much for my positive spin. I walk on, aware that my mind is about to slip into the abyss of populist governance and referenda. That way madness lies. A silent cathartic rant achieving as much as it is ever likely to. I need to stop thinking and remember myself. My wellied feet flicking through the spray, the slump of my shoulders and nice steady breaths. The day has darkened slightly, yet the cloud has lifted. The thick grey swathes that blew across with the rain and suggested light where there wasn't have moved through, leaving a higher but blacker umbrella in their place. I take off my glasses again and reach deep for the driest piece of pocketed tissue. I can see again, and the timing is perfect. As the field edge curves back towards me I spot a business of birds about 50 yards ahead. The mixed buntings that have been here all winter, up and down from the barbed wire of the fence and trail of bramble that hangs around it. I approach to perhaps 20 yards before stopping and lifting my binoculars.

There are probably 200 birds in total, although they splinter and scatter in smaller groups making a true count impossible, and the majority are yellowhammers. There are one or two males who are moulting into their summer gold, but the majority remain brown like the stubbled ground in which they work, lightly striped backs and russet rumps. Always a hint of yellow, but vague like the leaves of beech at summer's end that have lost their green rather than gained any autumn ochre.

I have scanned this flock plenty of times this winter, prompted in part by the presence of a couple of cirl buntings a few miles away (close to where I was looking for adders). We sit right on the edge of the cirl buntings' range, but they are a species steadily increasing. The British population is almost exclusively found in southern Devon, although it was once widespread across the south. The decline through the twentieth century is linked with changes in agricultural practice, typical of many farmland species, but it is with the cooperation of farmers that the recovery in

Devon has been so successful, with 1,000 pairs now present. The two birds that are on the coast are hanging out with a smaller flock of yellowhammers, which has prompted my added interest in this group of birds. In truth, though, I am not sure I have the confidence in my own ability to discern a juvenile cirl from a juvenile yellowhammer. The adult males, with their black masks and eye stripes, are distinctive enough, but spotting the subtle variables among a mixed flock is tricky – especially when they won't stay still. There is a smatter of reed buntings, though, the pale moustache a giveaway of both sexes, and a couple of bigger, bulkier corn buntings. The reed buntings will disperse elsewhere to breed, but the corn buntings have established themselves here in the past four or five years. I will pay them greater heed another time, when the weather is better and they are in voice. Right now, I have a need to scan the ground and fence posts for something else.

I sometimes wonder how my love of birds of prey was cemented. It certainly isn't linked to some deep-rooted bloodlust or fascination with the macabre. I can appreciate the process of a bird hunting, the puppet-on-a-string float of a harrier or arrowed rip of a peregrine falcon, but take no pleasure in seeing something else killed. I wasn't untypical as a child in the countryside, although my morbidity was limited mainly to insects: collecting a squadron of red ants to drop into a nest of black ants before watching the ensuing battle, or catching flies and throwing them into the web of a spider. I got some enjoyment from the family ratting efforts, my dad digging out a nest in the chicken coop as my brother and I stood and waited with stout lumps of wood. It was the only time Jason, the most placid golden retriever in the world, showed any kind of aggression to something else living, and I was happier to let the dog do the dispatching rather than swing a club myself. One of the pleasures of angling is putting the fish back and seeing it swim away, a contradiction I struggle to explain to myself let alone others. But essentially, the thing that set raptors apart from other birds in my esteem was nothing to do with their methods of food collection.

Instead, I think I was struck by their scarcity. An association with the wilder places – the moors, mountains and rugged cliffs. The kestrel was the only bird of prey I ever saw with regularity as a child and glimpses of anything more unusual would come only when on holiday, save those once-a-year flashes closer to home. The fortune of many of our raptor species has improved in recent decades, a sad exception being the kestrel that was so familiar to my young self. Greater protection and the banning of chemicals such as DDT have seen species like the buzzard, red kite and peregrine thrive. Always, though, their number is finite. Their place at the top of the food pyramid is the most vulnerable seat of all, a survival dependent upon every rung of the ladder beneath remaining intact. And success is relative. Much is celebrated of the enduring stability of golden eagles in Scotland, but 500 breeding pairs does not amount to many once you spread them across all those mountains and glens.

Our smallest raptor, the merlin, is not faring quite so well as others. This small, dashing falcon is merely hanging on across much of its traditional British range, its size perhaps a factor in its susceptibility to pesticide use. As traditional ground-nesting birds, in the UK at least, they are also prone to predation from mammals such as foxes or badgers, but an increasing number are beginning to nest in trees which may help them to adapt to changes in land use and climate change. Even so, the breeding population in the British Isles is probably fewer than 1,000 pairs, and they require considerably less territory than the golden eagle. In the winter months comes an influx from further north, with many birds arriving from Iceland. They will often follow flocks of smaller migrants such as meadow pipits, sticking close to a familiar source of food.

Early in our second winter here, I was driving through the low cloud at the top of the hill to the west of the cottage and disturbed a merlin that was sitting on a fence post. The view was brief, the bird taking flight and ghosting into the fog almost at the moment I saw it, but my eyes took a snapshot that I can almost recall now.

Then, in brighter conditions, I got a better view, as she (it was a young female) sat slightly hunched on a thin tendril of hawthorn. I was driving again and slowed to a roll but she stayed put as if oblivious of my threat, her eyes distracted. She remained until the spring, delighting Sue, for whom the merlin is a favourite bird, and on one occasion remaining still on a fence post as we inched past in the car with the windows down. If I walked up the hill then she would take flight while I was still at a distance, yet the car seemed to render me invisible, save that first encounter.

She returned the following autumn, frequenting the same perches and looking a little more wary, but surely the same bird. This time around, though, she seemed to have a fall-out with the local hen sparrowhawk, and the two engaged in regular aerial fisticuffs. The feud seemed to have an added edge in that it wasn't the usual raptor ding-dong that one might witness. There was no territory issue, or squabble over food, more a look that the sparrow-hawk, being a fair bit bigger, rather fancied making food of the merlin. And while I told myself that the merlin tired of the tussle and moved elsewhere, I rather think that the outcome was such.

I have made odd sightings of merlin locally since, but always in the autumn or early spring, birds moving through. A merlin is distinct in flight, dashing and compact, especially when hunting. They fold back their wings, almost cupped around their bodies between swept-back bursts. I see a similarity with a mistle thrush in flight, but the merlin is sharper and faster, and at the point where the wings tuck around the body they briefly dissolve into nothing. A clever cloak unless you are a pipit caught unaware. This winter, we have had another bird overwinter. A favoured haunt is along the line of fence posts that breaks the horizon on the ridge opposite, in view, albeit from a distance, from the comfort of the sofa. Interestingly, though, despite the proximity to us, sightings haven't been especially frequent and I rather think that is due to a change in our habit. The windows through which we stared non-stop in the first couple of years of living here have become slightly too familiar. Old habits, notably the television or the computer on which I type, have taken our attention, and I might not have noticed the merlin more than once or twice were it not for another birdwatcher. He has been walking the Roman road once a fortnight and invariably sees not just one merlin, but two. He has seen both birds together – and aged them – and without his success I might have gone through the winter unaware. It is odd, though, that even with the knowledge that two merlins have been overwintering within a few wing-flaps of home, I have yet to see both on the same day or see either to know for certain which I am looking at. And whenever I have walked this circuit, and this piece of road as the other birder has done, I have not had a single sighting. Instead, every glimpse has been from the lounge or the car, and while I am all for birdwatching in comfort, it does seem a puzzle. Perhaps, much as I had stopped looking outside as much as I used too, it is because I have stopped looking at all. In taking this place slightly for granted I have fallen out of tune with its rhythm. That awareness, deep within our unexplained sense, of a bird or animal that we cannot smell, see or hear, and yet know instinctively is there, is lost to the trappings of technology and artificial light. The other birder, coming away from those things, is looking with more than just his eyes, and not simply because he has purpose.

I have spent some minutes scanning today but seen nothing, although the nasal chatter of the buntings has steadily increased as my presence has become more benign. The sky is blackening deeper in the west though; a more significant pulse of rain might be coming, although it looks as though the wind might carry it off to the north. The same wind that in the next few weeks will be carrying off the meadow pipits and golden plover, with the merlins shadowing them as they go. In their place will come the swallows and martins, chaperoned by a small falcon of their own – the hobby. I don't expect to see any swallows just yet, although my earliest ever sightings (two swallows and a sand martin) came on 11 March, and today is the 10th. The wind is coming from the south-west, and bringing the mild, if damp, air with it, but as this system works across the country so it will bring colder air from the north, keeping most birds just where they are. Migration is a risky business without the increasingly erratic weather that comes with a fluctuating jet stream, and even as our winters grow milder there can still be a sting in the tail.

Three winters ago, the mild conditions fooled many, with blackthorn beginning to blossom and the green of ramsons filling the verges. Then came a beastly Siberian blast carried on a wind that was relentlessly cold. We were cut off from the world for several days, and the snow drifts on Eggardon were reckoned to be up to 15 feet deep.

It was the abrupt change in weather that was most damaging. The weather forecasters knew it was coming and we, the people, were able to prepare, but in the preceding weeks there was no hint of what was to come. There was quite a change to the cast of birds that frequent our garden, and an almost complete abandonment of wariness. The apple halves that we put out for the blackbirds became highly contested. Thrushes of several species bickered for every beakful, but the fieldfare were the most dominant. One particular individual became increasingly possessive, standing guard over a fruit and seeing off anyone who dared sneak a peck.

When survival is so delicately balanced, many natural instincts are completely forsaken. A pair of reed buntings, and a smatter of yellowhammers, birds that might visit the garden but with a tentative approach, would wait on the steps outside the patio doors and hop around my feet as I scrunched out to feed them. None of them could risk foraging any further afield for food. Instead, for two or three days, the normal boundaries within the natural world blurred, until, with the weather settling and the initial shock subsiding, everyone shook out their feathers and remembered their place.

The thaw came first on the coast, and once the snowplough had cut a route out of the village it was there I headed. I love wrapping up warm and walking in the snow, but I also needed a reassurance that winter itself was drawing to a close. I wanted to see a celandine and perhaps, given that the sun was doing what it could to break the grip of the cold, an adder.

I headed to Chesil, slightly east of where I searched for adders a fortnight since, and took a path into the nature reserve behind the beach. The ground, normally boot-squelchingly boggy, had been hardened by the frost. Between the clumps of club-rush, which I would normally use as stepping-stones to cross the marshland, shone slivers of ice drawn up from the ground beneath and blast-frozen by the wind. Here and there the ice would crack underfoot, but the bog beneath retained a solidity, giving slightly if I dawdled, but nothing like the peaty slurp that I would normally find. This patch is normally thick with snipe in the winter, and rather like the adder spot inland, I avoid the location as a result. I once inadvertently flushed a couple of dozen, but with nowhere to probe, my footsteps had nothing to disturb. Instead, the bird sightings had a distinctly macabre edge. I first noticed a skull with no sign of a body. The bone was picked clean, but from the back of the head two long black feathers plumed. It was a lapwing and, despite its condition, I don't think it had been dead for particularly long. In those conditions, a source of food would be shared by more than the expected scavengers. Alongside the gulls, corvids and foxes would be blackbirds, wood mice and possibly even lapwings themselves. It seemed that a wintering flock had been pushed out of the water meadows and ploughed fields where they had spent much of the winter and squeezed towards the coast. There, the sea presented a barrier too great. There was not the food or shelter on the land, and the flock had not the energy to cross the water. They could do nothing but sit and starve.

There were a dozen or so bodies scattered across an area the size of a tennis court, and this pattern was replicated up and down the coast with golden plover perishing alongside the lapwing – providing a food source for others at least.

We might yet get a snap this month, a sustained Arctic blast, but nothing like as severe or prolonged. And as I walk on, past the three masts and the burial chamber, I find myself slipping off my jacket to stroll in short sleeves. That threatening wedge of black has been hurried on by the wind like a parent ushering a child past a toy-shop window. It didn't manage to spill a drop here and, despite the steady movement, the air feels almost close. Perhaps I should have pushed myself a little harder and fished today. Wind does not make angling easy, but the ripple it cuts can diffract the light and make fish more confident in clear water. The urge isn't quite there, though. I will fish on Sunday, the final day of the season, but for now my head remains in a slightly different place.

Anniversaries generally pass without too much reflection on my part. I like to mark moments, and give thought and appreciation, but I am not a person who makes resolutions or pains as another digit is added to my age. The pandemic has made everyone a little more aware, though, and some people have suffered deeply from the isolation and anxiety. A few thoughts hit me this week, however, as I turned 47. It is just a number, and ageing is not something to fear, yet I found myself pondering parenthood and the reality that it is something I will not experience. A dream likely prompted the maudlin. A knot that my subconscious cannot unpick without some conscious influence. I can rationalise most things to myself, even if I do not openly admit my failings or then act upon such process. But instinct is a tough impulse with which to reason. I can look at the practicalities of Sue and me becoming parents, through whatever means, and every shred of logic suggests it to be an impossible path. Yet that doesn't redress the issue of what I, and we, have always taken for granted. Parenthood has always been something that is coming in the future, for me until I met Sue, and then for us in the twenty-three years since. It has been an inevitability – a step in our process. A thing that until very recently has always remained reassuringly certain. There will be a way around whatever has hindered up to now.

Except for time. The passing of it. There might be routes around physical obstacles, but even they are unrealistic. And while I can ponder within a settled mind and clearly accept the way of things and the positives that might present themselves as a result, my subconscious, which is built on things more complicated than logic and reason, cannot come to terms with it. So it is that 47 feels slightly more than a number. It is a reminder of what will not be. And whereas the majority of the things that my teenage self would have expected of my middle-aged self have not transpired, they can also be levied against those things that have. Except for this.

I feel I should allow myself some pity, although it doesn't fit with my internal coping mechanisms to do so. In the past I have used guilt as an extreme short-term solution, particularly when vulnerable to the darkest of thoughts in my late teens and early 20s. Forcing myself to think of the impact my actions would have on others, of the guilt that they might then feel. The horrible notion of people blaming themselves or feeling a responsibility for something they had no control over. It was a tactic that worked, obviously, although I have always been fortunate to have so many people in my life that I could visualise in those moments. But not being a parent is not a thing of shame or contrition. It is okay to feel disappointment and upset without telling myself that other people have it worse. Of course they do – there will always be someone in a worse situation than your own – but that shouldn't prevent a moment or two of self-indulgence.

And it should be said that there are also positive aspects to not being a parent. The glorious lack of responsibility and the freedom to be. In fact, the greatest benefit is right here and now. Not simply that I wanted a walk and so went, but that I am able to step out of time when I need to. I could guess, quite accurately, at the hour, but it is irrelevant. I have taken what I have needed and done so without impacting anyone else. The answer to my slight concern of turning 47 has been found by stepping out of the concept of time and its passing. By busting out of the institution and allowing a moment to last as long as it needs to.

I have another thought as I wander down the hill towards home, as I ponder Pliny's words on mistletoe. If a Gaulish age is thirty years long, then I am only midway through my second. And if the Coligny calendar points to cycles equivalent to sixty months then I am only 9 (and a bit) years old.

They really are only numbers.

4

THE GROWING MOON

A Sensitive Plant in a garden grew.

(Percy Bysshe Shelley, ‘The Sensitive Plant’)

THERE IS A FASCINATING passage in Gilbert White's *Natural History of Selborne*, contained in a letter to zoologist and regular correspondent Thomas Pennant, dated 17 August 1768. White has distinguished, 'past dispute', the three separate species of willow-wren. The familiar assumption of the time was that the willow warbler, chiffchaff and wood warbler (as they are all now known) were a singular species and classified as such, but with a specimen of each at his disposal, White determined otherwise. His observations are very much familiar to birdwatchers today – the slightly different sizes, the colour of the legs, the variations and sharpness of the plumage – and, of course, the songs. White notes that the largest willow-wren 'haunts only the tops of trees in high beechen woods, and makes a sibilous grasshopper-like noise, now and then, at short intervals, shivering a little with its wings when it sings', a description that points perfectly to a wood warbler. An earlier letter refers to a smaller willow-wren, 'the chirper', who is the 'first summer-bird of passage that is heard' and 'begins his two notes in the middle of March'. This, quite clearly, is the chiffchaff.

White was not alone with his interest in the natural world, and he references others, such as William Derham and John Ray with whom he shared and challenged opinions, but his observations are significant. Not least in demonstrating how little study was previously made within natural history. It shouldn't necessarily be surprising – after all, education and exploration were unaffordable for most and remain, to some extent, benefits of privilege. There was no means of sharing information in the way that we do today. People travelled less and local information and beliefs remained within the dialects in which they were spoken – hence the existence of colly birds in the West Country. It wouldn't have mattered that a willow-wren was actually three different species because such knowledge would have had no bearing on life. More important was to learn which berries and mushrooms were safe to eat and which would kill you, or that ash and hawthorn might burn when green but flint should never be used to line the fire. Survival mattered above anything, yet humans also lived a lot closer to nature and understood better their place within it. A relationship that has foundered as we have advanced, industry and over-population squeezing too much from too little. Yet there seems to be an increasing interest within what is around us, deepened unquestionably by the pandemic and lockdown. The sudden shock of change and restriction has stripped back our needs. We cannot go to the cinema or the pub, meet friends for coffee or splurge on retail therapy. We have been forced to re-evaluate and suddenly we see what is under our feet.

Sorry, I digress, where was I? Ah yes – I was thinking about Gilbert White because I was pondering willow-wrens, and I was pondering willow-wrens because one has just sent me into a spin. And the silly thing is that I came here expecting to hear one. Ten minutes ago, when I started the car engine, my mind and ears were full of birdsong, but then the radio clicked on and I didn't have the sense to switch it off. The relentless reporting of the pandemic is draining in itself, but the denial, misdirection and political playing that unfold around it are truly abhorrent. I rolled the final few yards in a steadily bubbling foment, which, in many ways, made the moment that followed more powerful. The 'chirpers' have been chiff-chaffing for several weeks and from every direction. It is a lovely sound, but also one that is familiar. The 'sibilous shivering' of a wood warbler is a sound I have not heard for a decade, and never in this area. I did briefly see a wood warbler in a small coppice close to our cottage some springs ago and a local birder, Alan Barrett, found a singing bird on this common on one day of last spring, but they are a bird in decline and reduced range, certainly compared to in Gilbert White's day. The third willow-wren is the one I hoped to hear today, although I did not quite expect to be serenaded as soon as I switched off the engine and swung open the car door. But there it is, the slightly hurried and non-descript smudge of opening notes that then tumble into a glorious fading descent. The song of a willow warbler is unassuming to a point of apology. 'I'm here but I don't want to impose and I'll leave if it is inconvenient ...' An unobtrusiveness easily lost to the insistence of others unless your ear is sharpened to it.

And in an instant I have floated somewhere else. Not unlike the moment I drifted while listening to the skylark singing beside the sea. But whereas the skylark takes me to grassy downland and sunshine, the willow warbler carries me off to the west coast of Scotland where it doesn't matter what the weather is doing. Not that they are a bird uncommon elsewhere, although they are thicker in the west than the east, but my association is with a place where a soundtrack is most evocative. In part due to a holiday that followed an extraordinary late-spring storm. The wind blew in from the Atlantic so hard and with such intensity that the moisture was sucked from every leaf on every tree. They curled and fell and our arrival was met with a landscape stalked by an army of arboreal skeletons. A winter view lit by early summer. The bare branches remained busy with birds, although without the millions of caterpillars that perished with the leaves, they must have struggled to find a meal. Instead, the willow warblers sang their lament to the loss as they flitted and flicked through the burgundy of birch. They seemed so tiny despite their conspicuousness, a bird no heavier than a pound coin that had somehow flown from sub-Saharan Africa only to discover a very different aridity.

In the early spring of today, and by the time I have swapped trainers for wellington boots, my ears have found three different willow warblers in voice. It is unusual to hear any beside the car park and there are normally only half a dozen or so on the whole common. There was a substantial population here a few decades since – Alan, who heard the wood warbler last spring, told me of when he surveyed this place some twenty-five years ago and would count several dozen singing males before he'd even covered half of the site. As I stand and cup my ears, I can make out a couple more distant singers, suggesting that a group has dropped in overnight and will disperse when the mood takes them. There remain good numbers in Dorset, but the chiffchaff is far more numerous and they seem to be chirping from every tree. It is quite tricky to pick out individuals among the thrum of the morning, and closing my eyes is beneficial to this aim. Aside from the immersion is a chance to filter through the different layers. I do not have a particularly musical ear, tending to hear the whole unless I focus otherwise. Unless a bassline smacks me hard in the face, I have to actively pick it out, and to distinguish a trumpet from a French horn would require a significant amount of guesswork. The spring chorus used to wallop me like an orchestral wall of sound – I could enjoy it, and perhaps pick out one or two components, but the rest just melded into a single piece. I was keen to learn more, and identify every voice, but how on earth do you start? The problem was that, rather like not noticing the moon until it is full, I was not hearing the birdsong until it was at its spring peak. The key, therefore, was in listening to the entire introduction. From the winter song of the robin and two-tone pipe of the great tit; the whistle and trill of the wren; and the familiar repeat of the song thrush. Nailing down the dunnock was pivotal. His is a song that can be heard in December when the robin is his wistful self, but the tone deepens as the days lengthen, turning a rusty gate into a frictionless revolving door. A rich, fluty song that almost matches the blackbird and blackcap, and unless you know the source, it makes those other birds much more difficult to discern.

By mid-March, most of the native species have tuned up and are in strong voice, just in time for the additions that come from the south. Some of the warbler species remain a challenge. I always like to see a garden warbler to know for sure I'm not confusing it with a blackcap – a problem that increases as the season draws on and the birds begin to mimic one another. But by listening as each layer was laid, I was able to hear whenever a new track was added. Then came a point where the experience changed altogether. There was no golden moment, but an awareness that

I was hearing a collection of different sounds rather than a single cacophony. And now I can sit on the steps with Sue, our ears full of noise, and yet mine will pick out a distant skylark or corn bunting that Sue cannot always pinpoint. Were I to be dropped straight into that moment, though, without the gradual acclimatisation, then my ears would rattle. When visiting a different habitat, the reedbeds of the Somerset Levels or an upland moor, then I still have to consciously break down the soundscape, but here, in my familiarity, I hear a reassuring soundtrack that seems suited to the rhythm of my own life.

The air is busy this morning. It is cold and crisp, with a light haze that is clearing as the sun inches higher. As I drove over Eggardon, the view west had a look of autumn. A thin mist hung like a dew-dropped spider's web, thickening here and there in the deeper combes, but never quite enough to obscure the landscape beneath. A month has passed since we had any recordable rainfall and following a scorching end to March, with temperatures similar to high summer, the high pressure that has sat still across the country began to pull light winds from the north. Easter weekend was one of contrast, in temperature at least, with a cold edge bristling in on the Monday following a Sunday where the beaches thronged. Lockdown rules remain in place but the promise of release has prompted greater movement, and while some people are bending the rules a little too far, it is perhaps forgivable to want to escape a high-rise for a day and sit on the pebbles of Chesil to stare at the sea.

Even places like this have been busy. I brought Sue down at Easter (without considering the date) to show her a melanistic adder that I had discovered and found a picnicking group just yards from where the snake had been showing. It seems reasonable for them to have found 'Powerstock Common Nature Reserve' on a map or online and thought it a suitable destination to lay a few blankets for lunch, but I felt I should alert them to the presence of adders, especially with small children playing among the thickened tangle. I didn't want to alarm them, and, in truth, my main concern was probably for the adder itself, but there was recoil at the mere mention of a snake and immediate moves made to head for home. Sue and I had a quick peek into the black adder's winter sanctuary and there he was, a small coil of scales that shimmered slightly blue in the sunlight. He is a stunning animal, with a distinct tiling of symmetrical white scales around the underside of the mouth and neck. And he lay just a few feet away from playing children with each party seemingly oblivious to the other.

There were no other cars here this morning, but force of habit prompts me to work the latch of the gate with a sleeved arm rather than naked hand. The frost remains thick in the shadow of the oaks to the left of the gateway but has long since vanished where the sun has touched it. There is a curious monotony to this weather – the still, the sunshine and the cold of the night. Early spring has probably brought more frosty dawns than the whole of the winter, and the pattern seems set for some time yet. As I walk, the impact of the weather is immediately apparent. Principally because nothing has changed. I haven't been visiting the common much this year, but for four weeks or more the season has stood stock-still. The early blossom on the blackthorn, the first leaves on the hazel – it is as if they have been frozen in a moment. And normally, given the sunshine, I would expect to hear a steady buzz of pollinators but they too have been frozen by the chill. Late in the day to stir and too little pollen to keep them stirred. Honey production is slow and, for once, it cannot be blamed on either Brexit or the pandemic.

These are the patterns of weather that are likely to be increasingly common as our climate changes. Five or six miles above us, the jet stream is a constantly moving flow of air, shifting west to east and sitting between temperature extremes that generate from the Arctic and the equator. For the most part its passage is smooth, an invisible river in the sky that might shorten and economise transatlantic air travel but only in one direction. The weather in the British Isles is reflective of the jet stream's position, but as those temperature extremes begin to widen, often sharply, so the stream will kink and shift, then settling in a new position for several weeks and bringing singular conditions to a climate renowned for its variability. April is normally a changeable month, full of those 'showers that bring May flowers', but if this pattern remains as predicted, then this particular April is likely to be one of the driest and coldest ever recorded.

The natural world is resilient, and certainly needs to be with all that humankind throws at it. Species will adapt to environmental shifts, cling on when there is no other option and then look to bounce back when conditions are more favourable. Conversely, people, in the Western world at least, have become consumed by convenience. We are so used to having what we need when we need it that we are hopeless at planning or predetermining. The weather now will have some impact on this summer and autumn's yield, and although the farmers will be painfully aware of that issue, the consumer will drift along in blissful ignorance. The problem is that we have become too complacent, taking for granted that everything will remain in the manner with which we are familiar. A little over a year ago, when the threat of Covid-19 became something tangible, there was panic and people began to hoard. Toilet rolls, pasta, flour, yeast – shelves were emptied and for no benefit. Within weeks there were issues in refuse collection, from the sheer amount of produce wasted because people had bought more than they could ever use. When faced with a genuine disruption many of us reverted to animal type. Selflessness and togetherness abandoned for self-preservation. Just like sparrows squabbling at the bird table – grab what you can before it's gone and fight for it if need be. The pandemic has brought the best and worst of human behaviour, the most unsavoury aspect being that in such a situation it really is a case of the strongest and most selfish who will suffer the least.

It is a reassurance of sorts, when considering the negative impact of humans as a species, that without the level of evolutionary intelligence that has enabled such influence upon the world around us, there would be no altruism. That we can care for those less fortunate or able is worthy of note, especially for someone like me who tends always towards the negative. Benevolence is a wonderful thing.

I walk on a short way to clear the overlook of oak and get a decent view of the eastern sky. It takes me some moments to find it among the blue and brightness of the sun, but it's there, the moon – just a day or two new. In truth it is little more than a faint wisp, like a defiant little smudge of cloud, but it would represent a significant point of the Celtic calendar. There is, of course, the caveat of presumption in my thinking, but the month that will begin in another few days is likely *Giamonios,* which is antipodal to *Samonios* from where almost all agree the calendar does begin. Xavier Delamarre linked the word *giamo* to 'winter', with a direct link to the Breton language, but the question does remain as to whether the month is linked to winter's beginning or end. Delamarre points to the following month, *Simiuisonios,* from which *uisonna* he links to spring. The complete translation, he suggests, could therefore be the 'second month of spring', which in turn would point to *Giamonios* denoting the end of winter.

While I find myself once more agreeing with Delamarre's logic, I do again wonder if I am so prompted as it fits my own agenda. So much of the literature I read is beyond my skillset. I am no academic and struggle to digest the language with which those minds write. It is no fault of the authors, but rather an aspect of my personality. My attention tends to drift the moment my mind struggles to comprehend and rather than open up and absorb I float off in distraction. It isn't unlike the sense I had in school, studying Shakespeare or T.S. Eliot, and being anxious by potential misinterpretation without understanding that meaning is subjective. Rather than express my own sense of feeling or merely enjoy words for the way they were shaped and moulded, I would fear getting something wrong and then reject the whole concept. We are all prone to leaning towards detail or evidence that supports our own beliefs; the difficult thing is recognising that trait.

But as I question my own motive, and whether I am being swayed by a predetermined judgement, I have another thought. I have since found pleasure in those poems and plays I pushed back against at school, because I have ceased to worry about correct interpretation. In as much as there is only one person who knows the true meaning and that is the author. The Celtic calendar, save an extraordinary archaeological discovery, will never be definitively explained. Much like the existence of ghosts or the infiniteness of space, some mysteries will be argued no matter how much research is undertaken. There is value, therefore, in my own sentiment, just so long as I feel justified within myself. These thoughts were bumping around as I walked to the Moon of Winds, and I mused then at the value of those voices that speak outside of traditional academia. There is weight to personal connection, and much as I have developed my own interest as I have empathised with those who once walked the paths I am taking, so there is surely credence within those people who remain immersed within the culture of Celtic tradition. If I can feel a faint sliver of affinity from simply standing upon the ramparts of a hill fort, then someone who has engaged deeper is likely to have insight that is built upon cognitive intuition rather than peer affirmation. As Barry Cunliffe noted in regard to Roman literature, we should not completely ignore a vision of contemporary or near-contemporary experience.

I have not walked particularly far. In fact, I'm more or less at the same point where I stopped to seek out the moon – which now, as I glance skyward once more, is veiled behind a thin wash of white cloud. A patch of gorse had grabbed my attention, a spot which can be good for a lizard or two. A slow sneak found nothing, though, probably in part to do with the broken sunshine. The air remains cold and the sun hadn't built up any great weight before the haze set in, and unlike adders, who will often sit it out with just the faintest hint of heat in the air, a lizard prefers less cryptic warmth. I reach down and rest the back of my hand on the ground and find no more warmth than in the air. There will be spots, I'm sure, where the first of this morning's rays has touched enough to encourage the scuttle of lizard feet. A deadened, dry log or high curl of last year's grass. The common is a good spot for viviparous lizards, with a healthy population of slow worms, too. On a good day I would expect to make double figures, with counts exceeding twenty if I concentrate hard and include the 'heard-only' scuttlers in my tally.

The gorse itself rather typifies the stilted spring. The green has limed from the emerald of mid-March and the flowers are shrivelled and slightly bleached. There is, unsurprisingly, not the slightest hint of coconut, even when I stick my nose into the thick and draw long, slow sniffs. There is a tickle in my nostril but prompted by the chill of the air rather than the pollen of a flower or prickle of spikes. There are three species of gorse in the UK and this, I presume, is common gorse (*Ulex europaeus*). The fact that the flowers are at eye level would certainly rule out the dwarf species and the ubiquitous nature of common gorse would also lead my thoughts to that end. Gorse is unusual in that it will flower at any point in the year. The sharp yellow petals are most likely to be seen in spring and early summer, but they may also provide splashes of colour in a deadened winter landscape. It is a risk worth taking – there are plenty of pollenating insects stirred by mild air or those, such as honey bees, that may reduce in number but do not fully hibernate. A yellow flower might hold quite the monopoly on a January afternoon.

To be honest, today is beginning to feel more like January than April. I left my fingerless gloves in the car and I am almost tempted to go back and fetch them. With the sun now masked, the northerly breeze is delivering a decent bite despite its lightness, but the landscape certainly looks more akin to spring. In the wider scrub behind the gorse patch loiter a couple of huddles of silver birch, the rhubarb and custard blush of the catkins cutting contrast against the fresh green of the unfurling leaves and the peeling bandage of the trunks. The oaks are still stuck in late winter, but the smaller trees that edge the open, the hazel and thorns, are looking almost as they should. The curious sense, though, is that of standstill, although having had that early-March blast, everything is probably now where it should be at this moment.

I walk on and, upon reaching a fork, resist my usual route. I normally head neither left nor right but straight ahead, into an uneven triangle that is managed for marsh fritillaries but to the benefit of adders. This is where I found the melanistic snake and where he will almost certainly still be today, but my footprints, and those of Sue and my parents (with whom I shared news of the adder), have left a fairly noticeable impression and I don't want to draw too much attention. So, instead, I veer left to follow the track up through the centre of the common, trying to avoid the scrunch of the loose gravel so I can keep my ears tuned. The short run of gorse and bramble on the right has homed a pair of stonechats for the past few years but today there is no sign. The male should be fairly conspicuous, a slightly restless flit through the cover as I pass – nasal clicks and 'tchiks' and agitated tail flicks.

The common has always felt as though it should support greater variety of small birds. It is managed by the local wildlife trust, but sympathetically, and although the scrubland is fairly isolated in the wider roll of rugged pasture, it is also extensive. The trees are always busy, with lots of nuthatch, treecreeper and great spotted woodpecker. There are marsh tits, goldcrests and bullfinches, yet the birds of more open country are thin. Plenty of whitethroat will arrive in the next few weeks but surprisingly few remain to nest, and larks and chats number less. It could be representative of shifting populations, influenced in part by climate change. Along with the vast numbers of willow warbler that Alan Barrett used to count was a pocket of willow tits, a species now absent from Dorset. That they clung on at Powerstock Common, while vanishing from much of their traditional range, might point to the suitability of this place as a habitat, the willow warblers also shifting geographically. Yet the relative lack of linnets, yellowhammers and stonechats doesn't necessarily bear this out.

One thought, based on personal observation rather than proven evidence, is the presence of ticks. There is an extraordinary number of ticks in the undergrowth of the common, to the point of unpleasantness. Most, I presume, are either sheep or black-legged ticks (which are often hosted by deer), although that is based on probability in terms of frequency and I have little interest in studying them long enough to know for sure, tending to brush them off as soon as I spot one. I will find them inching up a trouser leg in the thick of winter, whereas in early summer, when the larvae have first hatched, one single push through the long grass might deposit fifty or more on each knee. For a couple of years, I conducted reptile surveys here as a volunteer for the wildlife trust, stopping every few paces to wipe the inevitable cling from my legs. It was laborious and unavoidable, the tins that I would turn to count beneath being tucked away from regular footfall. Even the reptiles themselves would play host; I found an adder with two ticks clinging to the edge of an eye, and lizards carrying more than twenty passengers that crammed into the folds of skin around the leg joints. Several lizards appeared moribund, alive but stupefied, unfazed even when I picked them up to move them from the threat of my boots. If the ticks could find chinks in the scaled skin of reptiles, how many might attach to a barely feathered nestling? Those birds of open country that so often nest on or very close to the ground could see their offspring overwhelmed.

I can vouch for the impact ticks can have on health. As meticulous as I was at checking for unwanted hitchhikers following a survey, one or two would invariably find a fold and dig in. I wouldn't feel the fresh bites, but the old wounds, some of them scars from a couple of months since, would begin to irritate as my body recognised the incursion. I would pick up dozens through the summer months and the marks left were taking longer and longer to heal. Eventually, my health took a more distinct dip. Aching joints and clammy skin met with nausea and exhaustion. Blood tests were inconclusive but a diagnosis of Lyme disease saw me on a long course of antibiotics and a mental aversion to a place that

I had come to love. Having once visited several times each week, I all but stopped going to the common altogether.

The gradient is smooth and none too taxing, just a gentle rise to the highest point of this route. The coppice on my left has tapered and is thin enough now to see through the trees to the open square behind. There the yellows of last year's fade are punctured by the clean green of growth. Coiled fronds of bracken poised to break and, if the day was warmer, another smell to remind me of spring. Bracken has an earthy scent, almost fungal and nothing like the sweet fragrance of gorse, but it takes me off to childhood holidays to Dartmoor, of high-hedged lanes and stream-side glades.

I slow briefly to give attention to a pair of cedars that outreach the rest of the treeline. They house a raven nest and the occupants are usually quite vocal, although nest-side conversation generally consists of plinks and clicks rather than the loud cronks of flight. There is silence today, but the avian soundtrack is otherwise apt for this Moon of Growing. To my right, chiffchaffs and willow warblers seem to spill song from every birch, a blackbird flutes from one of the young oaks that dot the scrabble, and further on, beyond the brow, is the gentle retire of a mistle thrush.

This year has thus far been full of mistle thrush song. From the very beginning, before the birth of the Quiet Moon, the soft voice filled our cottage morning. It is similar to the blackbird but lacks the secondary flurry and is more hurried and less demanding. Sharp contrast to the abrasive 'clack-clack-clack' call that accompanies a mistle thrush in fleeting flight. It feels as though I have heard the song wherever I have walked, with several birds often sharing the air. They might not all be staying to nest, though – it is possible that they are winter visitors stirred by instinct to sing.

A couple of years ago I walked from home and heard a song on the wind that I could not fully discern. I wondered at first if it was a corn bunting, the resonance seeming to match the bunting's jangle, but as I got closer and cupped my ears I soon discarded that notion. Instead, the call had a curious familiarity that I couldn't place and was more like a woodlark in tone. I knew that I would have to physically see it in order to identify it, and so followed my ears, negotiating a barbed-wire fence in the process (I'm sure the farmer would have sympathised with my curiosity). It came as some surprise when I set my eyes upon it. A redwing, a thrush that spends the winter here but, save a handful of pairs in the far north of Scotland, does not breed and therefore rarely breaks into song. But this individual had been stirred by the lengthening days and a small injection of testosterone. Leaving me a puzzle before he flew north.

I have passed the brow and begun the downhill scamper beyond, and someone has flipped the avian soundtrack over to side B. Gone completely are the willow-wrens, replaced by tits, song thrush and blackbird, almost all of which are coming from the great beech towers to my left. A nuthatch bubbles, but otherwise the noise has considerably softened. The area to my right is a thick of squelch, with rush and tufted grass wobbling on the surface of a great brown blancmange. I once made the mistake of trying to cross it, having dived into the trees on an autumn search for mushrooms, and repeatedly had my wellingtons slurped off my feet. I explore far less now, in part because of the ticks, but also, at this time of year especially, so as to leave the wildlife be.

On one early visit, I followed a deer track into the denser woodland to the south. It was early autumn and we had only lived in Dorset for a couple of months, but I was exploring whenever I was able. The day was thick and overcast with not a breath of wind and the canopy offered no respite, just a deeper stifle. The deer track fizzled into nothing and the ground began to soften, so I sought a route out and picked up a more substantial path. I walked for another ten minutes or so before the trees began to scatter and the world opened up. There I headed up a small knoll to have a scan around. Looking north, I could see a wedge of land that looked incredibly like Eggardon Hill and through binoculars I could see that it too was fortified. I turned to look south to see Eggardon itself but found nothing other than an unfamiliar roll. As I turned north once again, I realised that I was looking at Eggardon Hill and in that moment the whole world seemed to swing through 180 degrees. My inbuilt compass, normally so reliable (in the countryside at least – drop me into central London and I am lost), had failed, and, rather like a needle pulled out of place by a magnet, my whole self had to bounce about and realign. I was left in a spin, a dizziness not unlike motion sickness, and it took me some time to walk it off.

I have stuck to the paths today, the steady decline having brought me down to the old railway line which offers a comfortable and flat route back towards the car. Here is where Alan heard the wood warbler last spring but there are none singing today, although the hazel-edged copse to the north seems to be a single, solid chirp of chiffchaffs. This was once the Bridport Railway, a branch line linking Bridport and West Bay with the Heart of Wessex line at Maiden Newton. Like so many rural links, this stretch of railway was marked for closure in the Beeching reports of the mid-1960s, but opposition, substantiated by the narrow lanes and lack of alternative routes, kept trains running until 1975.

The old track that remains has enormous value to wildlife, though, providing a corridor between habitats and colonies. Rather like the links that back on to the sea at Chesil, this thin strip allows movement and connection, just as it was originally constructed to do. Being slightly lower than the point where I began this walk and also being more tightly hemmed in by trees, I find my nose tingling for the first time today. There is a soft, slightly fungal must, edged with the acidic sharp of fresh nettle and the faint, sticky tang of 1,000 buds ready to burst. There is colour too, the yellow of lesser celandine looking so sharp against the lily-pad green of the leaves, and further into the trees the white of wood anemone, another member of the buttercup family. It is odd, but until the smell of spring gave me a walloped reminder, I hadn't paid much heed to the colour around my feet. I know that here will be fewer flowering plants along much of the route I have taken, but my preoccupation with sound had stopped me from looking hard for what might have been there. It is interesting, but until the fug of late winter has truly left me, I tend to engage with limited use of my senses. I might have days where I am fully responsive, but often, and especially when I have nudged myself out, reception tends to be fuzzy. In another few weeks, and likely by the time I take my next walk, all channels will be wide open, but for now I still have to work at it.

One thing entirely absent today, though, has been butterflies. I have yet to see a single one, and even in a daze I wouldn't miss one in flight. It remains too early for many species of course, but nevertheless I would have expected orange-tips and holly blues, with a good smatter of peacocks and commas. The cold has helped the overwinterers and early-emergers in no part, and despite the signs of growth and of spring, this will be a poor season for many species.

I pause beside the railway hut, recently rebuilt having been razed by fire. The canopy is opening wide once more, the trees stepping back as the path – the railway line – enters a cutting that can be glorious for reptiles and butterflies, but equally so for ticks. The cloud is threatening to break, the glow of the sun strong enough to avert my gaze. There is a warmth there, too, and I imagine the adders and lizards will be stirring. And despite the frosty start and the enduring chill, there is a firm sense of what is to come. We are well past the equinox, the day pinching from either end of the night and the sun edging ever higher in its daily arc. 'Winter's end' – *Giamonios* – really does feel appropriate for the forthcoming moon and would have felt even more so 2,500 years ago when the winter was so much more marked and punishing. An alternative interpretation of *Giamonios* ('among many') has also struck me as seeming rather apt. 'Shoots-show' is the meaning derived from the Coligny calendar by Caitlin Matthews, a name I have noted as a source for several online lists of 'Celtic moons'. Matthews is, arguably, a contemporary voice to which I might pay proper heed. She being – and I'm sure she wouldn't object to the description – in touch with the spiritual aspects of Celtic society and tradition to a level deeper than many other experts in this field. It is easy to disregard certain research because it is anecdotal or lacks a level of required academic scrutiny, yet so much is only learned through such means. Unless we question what is accepted, our knowledge will stagnate.

It seems appropriate to mention Gilbert White once more in reference to that cause. Just yesterday Sue and I were discussing earthworms and their value to soil quality. I suggested it was Gilbert White who had unearthed their benefit, whereas Sue thought it was Charles Darwin. As it transpired, we were both correct, Darwin's 1881 book *The Formation of Vegetable Mould Through the Action of Worms* having been built upon White's observations. Both men, however, would have seen such thoughts met with scepticism or disinterest, particularly where they contradicted traditional belief. Even today, there are those unwilling to accept learning outside of the knowledge that they have accepted for a lifetime. In rural communities, plants such as ragwort are uprooted as a matter of course, there being some threat to livestock (especially horses) if ingested, but the unwavering reaction being one of revulsion and in eradication. Species such as crows and wood pigeon, rats and foxes are viewed as vermin with no 'worth' to humankind and might be systematically persecuted as a result. The justification, aside from the validity of protecting one's livelihood, being self-appointed managers of all species. Protecting songbirds by removing species that might predate nests rather than look to improve the habitat in which they live. Predation is an essential part of the evolutionary process and, while it is important to consider our own role within it, and our absence from a functioning food pyramid, we too often tend to play God and attribute human characteristics to the creatures that we malign.

It is difficult to alter an opinion, especially when it contradicts our personal values, but we owe it to the environment to do so, particularly from the seat of privilege from where we preside. And there is personal benefit to be gained too. By engaging as equals rather than overlords, we can draw a deeper sense of mental worth. There is once more a spirit of benevolence and a wonderfully reassuring level of insignificance. To allow one's self and actions to be inconsequential is to liberate us from the shackles of societal structure. We are atoms and dust, just like the wasps we swat and the ants we powder. This is how we truly grow, and to accept that is to feel free.

5

THE BRIGHT MOON

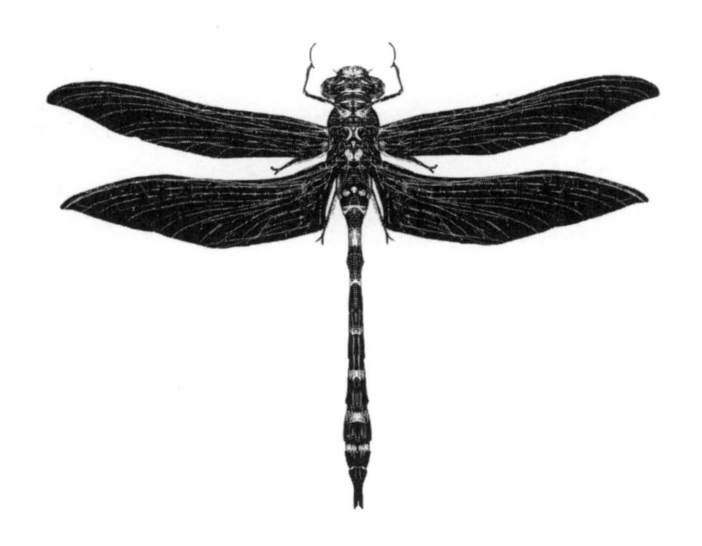

Again the violet of our early days
Drinks beauteous azure from the golden sun,
And kindles into fragrance at his blaze.

(Ebenezer Elliott, 'Spring')

THIS FEELS RATHER odd. After driving for almost an hour, further than I have travelled for at least fifteen months, I find myself in a large, gravelled car park, with log partitions and an area set apart for coaches. At one end is a toilet block, a coffee shop (which remains closed), an information hut and a picnic area. And there are people. Not many, but enough to make me feel slightly unnerved. I simply haven't been anywhere other than the supermarket for so long that I feel exposed to be in a public place once more. And not only that, I am meeting people here, too. Chris, Will and Matt are all due to join me and are all running late – not that I mind too much; it does give me a few moments to record a few thoughts and reflect upon the oddity of everything.

I am at Ham Wall, an RSPB reserve situated on the Avalon Marshes in Somerset, an area that historically has had a volatile relationship with the sea. These lands flooded as the ice receded following the last Ice Age, leaving a few isolated islands such as Glastonbury Tor and Brent Knoll. Such spots were settled and pathways constructed between them, the most famous being the Sweet Track which was named after Ray Sweet who uncovered the remains in 1970. Sweet was excavating peat, which had preserved the wooden pegs and planks that formed the pathway for nearly 6,000 years. The Sweet Track ran close to where I am now sitting, and its subsequent submersion reflected a period when the build-up of peat and reedbed reclaimed the area from the sea. However, the independent study group Climate Central have warned that, if unchecked, climate change and the subsequent rise of sea levels could result in this area disappearing once more beneath water by 2050. A somewhat sobering thought, particularly given the extraordinary wildlife that can currently be found here. The peat workings that once covered these flatlands have been transformed into a vast wetland paradise, and the response speaks for itself. I first came here a couple of years ago with my parents and, although they had been before and told me of what to expect, I still couldn't believe what I was seeing and hearing. I felt almost as though I was in a zoo, that there was no possible way that so much life could be in one place without having been brought in. The sense I felt was of surreality and I feel that same way this morning, although there is an added element today. Lockdown has left me, and many people, in a slight state of institutionalisation. Being out in public, alongside the prospect of meeting people I know and love, has left me curiously anxious. I have stepped out of a comfort zone that was not mine to create. A bit like a second or third date that is full of possibility and excitement but also represents a risk of pain or exposure. It is strange that enforced isolation can so quickly evolve from frustration to reassurance, although it is a feeling that resonates.

I spent some time in a secure unit when I was 20. Actually, 'secure unit' is something of a modern-day attenuation – it was a mental hospital, dating back to the First World War and since bulldozed and developed (for housing, obviously – nobody gets mentally ill any more).

'It's nothing like *One Flew Over the Cuckoo's Nest*,' I was assured as I arrived – they probably shouldn't have made that connection. As I lined up for my meds twice a day, in a shuffling queue of glazed eyes and blank faces, it was hard to compare it to anything else.

My admission was something of an inevitability. The previous months will remain blank in my memory, for the most part at least. I recall brief moments of lucidity, snapshots of awareness like waking briefly from a fever-driven delirium, and in which I could scarcely recognise myself. I couldn't sleep, or work, or be. I refer to it now as a breakdown, but that is largely for ease of explanation. A breakdown suggests something instant whereas my decline, while sharp when contrasted with the positive place I had been in, was a steady stumble rather than a dramatic fall. And I knew myself that I was unwell, to the extent that I had been willingly admitted to hospital, which also meant I could leave against the advice of the doctors. Which I did after just a week. In that time I'd been sent into the stratosphere with an assortment of drugs that would have made Syd Barrett smile, and it didn't matter that I couldn't be prescribed them outside of the hospital. I wouldn't need them – I was invincible.

The ensuing comedown was not so severe as to see me back where I began, but the world went very dark again. And there were often moments, alone in insomnia or alone in a crowded room, when I wished myself back into Park Prewitt Mental Hospital. Free from responsibility for myself or my actions, free from my own mind. Comfortably numb as I watched ripples pulse through the paint on the walls.

Introversion is as inevitable as it is unhelpful for those suffering from mental illness. As is an occasional reluctance to accept our own failings. I am not suggesting for a second that a week in hospital left me institutionalised, but it helped me to understand how it can happen. As our lives shrink back so they simplify. A routine smooths off the jagged edges, yet the less we have to do, the less we are inclined to do. Those things that we do without thinking suddenly require quite high levels of motivation, and the prospect of a 35-mile drive this morning felt as though I were journeying to the edge of the earth.

It was utterly painless, of course. I might not have the endurance I had as a white-van man driving up to 500 miles in a single day, but even as a Covid recluse this morning's journey caused no concern. The only issue came as I headed north from Yeovil and edged across the eastern edge of the Levels, as great drops of water began to blotch across the windscreen before hammering hard. Rain was not part of the forecast, not this far south at least, and would rather scupper today's objectives. I visited around this time last year and a highlight, and the spectacle that I teased Chris, Will and Matt with, was the extraordinary number of hobbies that filled the sky. When I was 10, I had a poster on my bedroom wall that featured a hobby, a photograph accompanied by the sobering inscription 'Fewer than 100 pairs nest in the UK each summer'. They were a bird almost as out of reach to my childhood self as a golden eagle: a dashing, exotic acrobat that was very particular about where it lived. We didn't live far from the heathlands on which it might be found, but far too distant for me to visit without the cooperation of busy parents.

On our visit last spring, though, we found the skies full of falcons. They wheeled and hawked in every direction, feasting on the abundance of insects that rose from the marshes. It would, I promised my friends a few days ago, blow their minds too, and a dry day was promised – perfect for insect emergence and perfect conditions for a hobby to fly in. Then came the rain. It persisted as I trundled past Dundon Beacon and Combe Hill, sites of hill forts and earthworkings created at a similar time to the ramparts of Eggardon Hill. It was a road I had not travelled before and an area that I was able to appreciate regardless of the splatter. The hills created a landscape as if drawn by a child. Perfect, wooded domes that bulged in perfect contrast to the surrounding flatness. Then came the climb up the ridge that marks the eastern end of the Polden Hills and, joyfully, the sky to the north was bright and broken. The cloud that chased me from home was checked by the natural form of the landscape. Our walk would be dry and a feast of falcons feasible.

It is difficult to track invertebrate numbers due to their small size, their relative profligacy and the difficulties of identification. And humankind has long viewed many invertebrate species as pests that we care little for and have no worth to us. We all know that 'bumblebees' are in decline, yet there are twenty individual species in the UK and some will be faring better than others. Butterflies are possibly the easiest insects to count and record, due to their visual appearance and habitat dependence. With fifty-seven resident and two regular migrant species in the British Isles, they are easier to track, and in 2015 the *State of the UK's Butterflies* report determined that, in the previous forty years, 76 per cent of our butterflies had declined in abundance and/or occurrence.

This rather sobering statistic is echoed across the wider *Insecta* class. In 2019, Francisco Sánchez-Bayo and Kris A.G. Wyckhuys published 'Worldwide Decline of the Entomofauna: A Review of Its Drivers', which found that globally more than 40 per cent of insect species are facing extinction. Their research points to habitat loss as the major driver of this decline, caused primarily by the increase of intensive agriculture, while the use of agro-chemical pollutants, presence of invasive species and climate change were also contributing factors.

An important element to consider in such research is the original base from which we measure. We tend, as humans, to accept as fact our first learning or understanding. My mind, for example, still assumes that there are nine planets in our solar system, it being the figure I first encountered and then had repeatedly reinforced as I journeyed through education. By the time that Pluto was reclassified as a dwarf planet, my interests and readings had moved on and I didn't have the inclination to change my default setting. It was buried too deep within my memory and had been recounted too many times. It has almost no bearing on my day-to-day life and none whatsoever upon anyone else's, but should the matter crop up for any reason then my mind begins with nine planets before a later memory recalls the fact that there are officially eight (with five dwarf planets). The amendment is applied rather than learned.

When I was a child the UK red kite population was restricted to just a handful of birds in mid-Wales. They were a true rarity. Today, aided by a reintroduction programme and a huge increase in that original Welsh stronghold, the kite is thriving. There were around 100 breeding pairs in the mid-1990s, whereas in 2020 there were estimated to be more than 10,000 birds in the British Isles. Such a dramatic increase might suggest a population explosion yet were we to compare the population today with that of 400 years ago, when kites were reported to be common scavengers in the streets of London, then the modern population might not be quite so impressive. Equally, were we to take the population today as a baseline, then the increase in a further twenty-five years will not be so marked. That initial burst has created a false impression of ubiquity.

This notion of a 'shifting baseline' was pondered by Ian McHarg in *Design with Nature* in 1969, and further considered by scientist Daniel Pauly in relation to fish populations. Pauly noted that many experts tended to base their ideas and theories upon the condition of a fishery at the start of their involvement, therefore masking the true depletion of stocks over a longer period. As a consequence, each generation begins with its own 'baseline' from which population trends are perceived. The true long-term pattern is very different to the one that we often consider.

It is interesting too that humans tend to consider such things within the constraints of our own mortality. We are reluctant to accept natural fluctuations of species because we ourselves, especially in the Western world, have evolved to overcome the impacts of disease, illness or disaster. So it is that we struggle to view situations beyond our own concept of time. A river, for example, is an ever-changing habitat, tens of thousands of years in the making, yet some anglers catastrophise about the increase in otter numbers, when any impact a naturally occurring predator will have on a habitat is localised and short term. Change is something we struggle to tolerate, though – we obsess with control but only because we ourselves have been so intent on eliminating any shocks and surprises that we, as a species, may encounter.

That shifting baseline only becomes truly apparent if we are able to look back to a time that preceded the formation of our individual measure. An impossibility, of course, although the Avalon Marshes offer a glimpse of what could be. This is a landscape given back to nature, but with a helping hand. Old peat workings tidied and flooded, the reedbeds planted and cut. It is nothing like it would have looked to the Celts who trod the wooden trackways and yet in many ways it is exactly the same. Certainly within a superficial sense, the endless swathes of reeds punctured by pioneering alder and pockets of scrub. The smell would most definitely be familiar, a deep must bubbled up through the water, hanging with the rot of millennia. A sweetness when left to mingle with the clean of the air but pungent and powerful when squelched up with a heavy foot. The sound, though, is probably the closest representation of this place as it once was. Not, perhaps, in terms of the tune – the cast will have changed slightly – but the volume will be much the same. And the sound is extraordinary.

It is strange to see faces again. After so long in solitude, actual social engagement is novel to the point of overwhelming. It is far from unpleasant – in fact we cannot stop laughing. It is particularly odd to see Chris, who I speak to via telephone on more days than I don't. His voice is so familiar that to see it being formed once again is surreal. For several minutes I have to let my eyes adjust and look away more than I would normally in order to ensure my senses don't begin to swim. I am reminded of a conversation I had some years ago with Zita, someone familiar to Chris and Will for most of her and Will's life. Zita's twin sister Jessie had returned from a trip to Australia, where she travelled for almost a year. The twins had maintained fibreoptic contact but, as they had spent their entire lives living so closely to one another, I was curious as to how it felt for them to be reunited after so long apart. Zita had been totally bowled over, but while the emotional jerk was quick to settle, the physical aspect took longer. It was some days before they were able to look at one another in the eyes without Zita feeling as though she might pass out. The sense was beyond dizziness, through nausea and into a situation where her whole self simply couldn't cope with the load. Having adjusted to life without the person with whom she had shared a womb and all things since, to suddenly find it again – to find her again – was simply too much.

It doesn't take me quite so long to adjust to seeing Chris's face once more, and once the initial giggle has settled into chat, the gaps of separation have bridged between us all. And as soon as we become less aware of ourselves, so we find ourselves struck by the noise around us. Just in the bushes that surround the car park we can hear blackbird, song thrush, blackcap, whitethroat, willow warbler, chiffchaff, robin and wren. More distant, for now, but steady and unrelenting for the rest of the day comes the call of a cuckoo, a sound that has vanished from so much of the lowland spring. This is nothing, though, I tell them; when we get out beside the scrub and reedbed, that is when things get really silly.

One of the most alarming aspects of the pandemic has been the pressure that it has placed upon our health service. The people working within the NHS have been incredible, although it takes a certain kind of incredible to do such jobs in the first place. I couldn't; I like to believe that I am a considerate soul – but I do not have the skillset and depth of altruism required. For many years I felt I might have value as a counsellor, my own issues with mental health offering an element of empathy for others. In truth, though, my motives were likely driven by the flicker of the illness itself, a curiosity of depression being that, as it forces introversion, so it might also confuse the ego. There were moments when I felt so alone, so detached from the rest of the world, that a part of me wondered if this was a burden borne of a higher purpose. Perhaps I was unique, I carried greater weight because I had more to unlock. In hindsight, I think I was experiencing a sense of self-preservation. In those really dark moments, when I was faced with something ultimate, I would use guilt to stop me from stopping, but that was not beneficial to my psyche. So there would come these moments when my mind would pull in the opposite direction. I would still feel utterly squashed, but somehow I could make myself a martyr to it – no one else could feel as I did because no one else had the capacity. On occasion, and invariably fuelled by alcohol, this would manifest as supreme confidence, bordering on arrogance. I might enjoy an explosive spark of connection with someone at a party on Saturday night only to re-present as a hopeless blob on the Monday.

This aspect of depression has probably been the most difficult to admit – particularly to myself. I am grateful, though, to have become aware and to have experienced my illness in the days preceding the internet. It is easy to see how people have tumbled inwards during lockdown – shut away within a spiral of mental diffraction. Exposed to those looking to exploit the vulnerable. Conspiracy theories and the deep state. There are rabbit holes far deeper than the one into which Alice fell, and for someone alone and desperate for answers, stumbling is easier than staying on one's feet.

The breeze is steady but the cloud slow to shift. There is no sign of rain, though, aside from the steady billow of willow seeds. A thick float of cotton wool that twists and lifts like a dance of mayfly. There are not the dragonfly numbers I had expected, and a word with a warden suggests that they are but a fraction of their usual May abundance, the cold of April slowing their emergence, although numbers are beginning to build. There would usually be vast squadrons of four-spotted chasers cruising the reedbeds and the dusk roost is apparently biblical. I was sent a couple of photographs by Will Snelling, in which the chasers mass on the reed stems like locusts. Thousands upon thousands of dragonflies, with black spots like half-opened eyes on the leading edge of the wings. Without the numbers of dragonflies, though, there might not be the hobbies I had so confidently promised. The hobby seems happy to stick to an insect diet in the early part of the year, twisting and tumbling and eating on the wing. Catching a small bird then requires perching and plucking, risking the attentions of bigger scavengers, crows or buzzards, who couldn't hope to pick the pocket of a hobby in the air. So it is that when the skies are full of flies, hobbies will often hawk alongside swallows, swifts and martins, feasting together in an uneasy truce. And dragonflies are the ideal prey for a hobby, bulky and protein-rich – perfect food efficiency.

Across the reedbeds the scene is just as promised. There is always a concern that memory distorts reality, but it is as if I underestimated Avalon. We stand, squat and lean in silence, eyes blinking, ears briefly muffled as we absorb the scene before us. We have only walked for a few minutes, a short stroll from the car park and along the path that edges the old canal. A constant flutter and song from every branch and a slightly bossy insistence on my part to ignore the first glimpses of water to our right in order to savour the view that has now opened up to our left. The wetland to the north is older and slightly less sculptured than the freshly dug lagoons to the south of us. It is all thick with life, but there is a deeper sense of connection to a place that has been allowed more time to establish its own balance. Mind you, there is so much going on before us that even I am surprised.

There are wider, deeper areas of water where great crested grebes and pochards dive and dabble, the grebes more lithe and agile, and striking in their summer dress. Rust-stained feathered jowls and inky plumes that splay like fans give the bird their name. Both sexes are similar in appearance, unlike the stouter, more compact pochard. The drakes are grey-bodied, with black tails and necks and heads coloured similarly to the jowls of the grebes. Through my binoculars I can see the red eye, curiously unassuming but likely due to it being lost to the colour that surrounds it. The females are more mottled and less coloured, broken browns that give them better camouflage on the nests that will be tucked into the edges of the reeds or marginal tangle. It is important to melt into the background when sitting duck here, for the air is full of threat, and a raptor I expected us to see alongside the hobby is already in view. In fact, there are two. Only slightly smaller than a buzzard, but with thinner wings and a longer tail, the marsh harrier is the master of the reeds, boasting a floating, effortless flight with moth-like flaps and paper-aeroplane glides. Fossil remains suggest that it was frequent here while the sea still had a salty claim on this landscape and this century it has made a dramatic return. Having almost disappeared as a British breeding species in the mid-twentieth century, the marsh harrier has expanded in range and habitat, although it remains most familiar in the places which give it its name. Here on the Levels, its recent resurgence reflects the extraordinary success of this environment, yet it is arguably not the most amazing avian story unfolding.

If we are told something enough times then we often create it as a memory even though it is not ours to own. Tales from infancy, when we amused others or embarrassed ourselves, can lodge as our own recall, particularly if there are physical snapshots – photographs or videos – to cement the image. My memories of the time that led up to my hospitalisation are thin. A sprinkle of blurs surfacing amid a deep swim of numb. Until the drugs shocked me sharply awake, the ward itself was desperately vague, but one memory is etched and it isn't my own. My mother recalls a visit when she and my father arrived to find me outside with my friend, Mike. We were leaning on a gate that overlooked a small paddock of green on the edge of the hospital grounds, smoke spiralling from our rollies and us in reflection rather than conversation. The moment obviously offered my mum some positive imagery at a time which must have been incredibly difficult, and I can remember it myself – but only through her eyes. What is definitely my own memory, though, is that sense of connection that smoking gate offered. Not from a nicotine hit (we were allowed to smoke in the ward) but with the outside world. The grass and dead nettle and dock stems, the crooked fingers of leafless branches and solid reassurance of rooted trunks. The important place. The cool of winter air and suspicious 'chack-chack' of a magpie. There wasn't much to see or feel but enough to feel that insignificance of self that is so vital at times of mental crisis.

Looking back at the dark times, those moments of connection sit like little islands in a vast, black sea. Sitting with my sister, Cath, on a late-summer evening watching the sun set over Crabtree Woods, near Winchester. I do not recall the going or leaving or conversation, but I know that Cath took me there for some respite and for a short time the veil lifted. I beached briefly upon another island beside the Dorset Stour in the company of my father and grandfather. It wasn't the fishing that lifted me, but a moment as we stood in the early morning looking across the meadows towards the river. The mist hung between the deep green of the grass at our feet and the autumn tinge of the treetops that edged the ridge on the far bank of the river. For several seconds my mind cleared and a glorious sense of ease swept through me like a shudder. That moment sustained me for the rest of the day and into the next, not by its recollection but more like when taking a gulp of air before sinking beneath the water's surface.

I mentioned a moon or two ago that the day would come when I wouldn't have to force myself out of the door for my own good. And rather like the disappearance of a nagging cough, the actual moment came and went without me noticing. It is only when the thought occurs as we stroll the boardwalk out to one of the hides of the marsh that I realise how easy I found the process of today – and it isn't just because I am meeting up with people I love. The summer solstice is less than a month away and the days are long and bright. There is an energy that seems to reflect in every living thing, an urgency to grow, to mate, to flower and to seed. But that urgency comes with less pressure. The stretch of the day, benign weather and plentiful food combine to deliver a feeling of positivity and production. There are lapwings lolloping and diving, whistles and 'pee-wits' to the bounce of black and white. A complete contrast to the desperate scene I found following the 'Beast from the East'. Plenty of threats remain, of course, both aerial and terrestrial, and lapwing eggs and nestlings will be eyed up by all manner of creatures with mouths of their own to feed, but this is a time of vibrance and being. Of new life and living.

The month of *Semiuisonns*. The 'second month' of or 'middle' of spring according to Xavier Delamarre's translation. Elsewhere, for this lunar cycle I have come across 'Mother's Moon', 'Flower Moon' and 'Time of Brightness'. The last of these is referred to by Caitlin Matthews in her book *The Celtic Tradition* and, having seen her name sourced for several lists, I decided to drop her a line. Caitlin's response was rather reassuring. As I earlier alluded, not all voices are treated equally in certain circles of learning, and while I concur to a point, this is an area where personal interpretation carries considerable weight. Not least in the sense that it cannot be wrong. Rather like 'The Love Song of J. Alfred Prufrock' by T.S. Eliot that I struggled with at school, there is actually no right or wrong answer, just so long as you can add validity to your reasoning. Caitlin pointed me to a series of papers on the Coligny calendar, and as I scanned down the list I recognised several of the names – Helen McKay, Catherine Swift and Garrett Olmsted. I had followed several of these paths before and taken snippets and thoughts from each, and although there was no solid corroboration between them, that is rather the point.

The Bright Moon feels appropriate to me. Not only in the strength of light, the sunshine, the colour and sounds, but in the way I feel at this time. I feel fresh and able; unladen and energised. It has been a gradual process but I am a different person to the one that stumbled around four moons ago. Relatively, of course. The vital thing, now, is to enjoy the moment.

The warbler count is up to eight. Reed, sedge, willow, Cetti's, garden, blackcap, whitethroat and chiffchaff. And the sound shifts as we step from scrub, to copse, to marsh. Among the sway and light clatter of the reeds the reed warbler dominates, a rather insistent and hurried 'tchirr' that is followed by a steady ramble of repeated whistles and clicks. The sedge warbler is similar in essence, but more nasal and insect-like to my ears and with greater variation in note. When I learned to distinguish them, the sedge warbler seemed far more plentiful, but the fortunes of the two have since shifted. Today I have only picked up a couple of sedge warbler songs but they can be hard to pinpoint from the all-round racket. They are at least two species of similar song that can be easily distinguished in appearance. The reed warbler is clean and sleek with a white chest, distinctive white throat and fairly uniform brown and burnt sienna back. The sedge is more streaked and upright, with a spoon-shaped tail and a very noticeable pale stripe above the eye.

Back on the pathways, from the willow and alder, sang two other warblers similar in voice but different in appearance. The blackcap is sometimes known as the 'nightingale of the north', although his song is, in all honesty, nothing like the bird to which he is being likened. It is fluty and loud, rather like a sudden burst of violin, and the garden warbler is very similar. I met a reserve warden last spring who had allowed me entry to record some of the few remaining nightingales in Dorset, and we chatted about how we distinguished the garden warblers and blackcap by song. The garden warbler is slightly deeper in tone, I argued, more flute-like and slightly more reminiscent of a blackbird. He laughed and told me he thought the very opposite. Rather like seeing colours, we can never be sure that we hear things in the same way as one another. It certainly helps to get a glimpse of a garden warbler before separating the song definitively from a blackcap and that is often part of the difficulty. Not here, though – territory is so prized in the arboreal veins that criss-cross the Levels that reticence might cost a pitch. Both species are frequently found singing within yards of one another and in full view; position yourself correctly and it might be possible to level them in stereo.

At dawn and dusk, the sound here will be even more impressive, and the orchestra would be complemented by a baritone addition – probably in the brass section. We have only heard one 'booming' bittern today, although were we here a month or so ago then there would have been a steady reverberation. And although we haven't seen a bittern yet, I certainly hope to – most likely a bird in flight. When not in the air, the mottled browns and slow, slow stalk allow the bittern to vanish among the reedbeds, impressive given its size. They were a bird that it was hoped the management of the Avalon Marshes might encourage, but no one could have dared dream of the dozens of males that now boom each spring. And that isn't all; a decade or so ago the great white egret was classed as a 'rare but annual' vagrant in Britain. Until recently I had only seen one in the UK and that was while mushroom hunting with my cousin Chris on the Isle of Purbeck. We passed a bird hide overlooking Little Sea on Studland and I commented on the fact that a great white egret had been drawing small crowds of birdwatchers each day. We popped into the hide, lifted a hatch and there it was, Chris deciding, quite reasonably, that finding rare birds was a lot easier than finding porcini mushrooms. On my first visit here, it was the presence of the egrets that gave the strongest sense of being in a zoo. A first, excited glimpse was followed by a second and third, before I realised they were everywhere. A bird that according to every bird book on my shelf should not be here at all was doing rather well for itself, following the pattern of colonisation set first by little egrets and also now by cattle egrets. But the speed of the expansion of these species on the Somerset Levels is the most extraordinary thing. Yes, you need the sheer expanse and the careful management. But create it correctly and nature will come. And it all starts at the bottom.

As an angler I noticed the fish as quickly as I saw the harriers and egrets. Not physical sightings but the tell-tale dimples, bubbles and billows of silt. The fish were thriving because the invertebrate life was thriving, and the insects, shrimp and daphnia were flourishing because the water was as it should be. Unfettered by excessive abstraction and the nitrate and phosphate imbalance caused from agricultural run-off. The reason that such a place can feel contrived is the sheer volume of top- and intermediate-level predators, yet that is clear proof of an ecosystem working precisely as it should. The bitterns, harriers, otters, grebes, cormorants and egrets are entirely dependent upon protein, and ensuring that the lowest points of the food chain were in place has allowed every step above to flourish.

It might look a little different and the pathways are certainly more stable underfoot, but this landscape is probably the closest to that of the time of the Celts, in terms of flora and fauna at least, that I might experience. And we are wandering the northern reaches of the Durotriges territory, the same people who walked these flatlands as were buffeted by the wind on Eggardon Hill. I like that thought and I think I understand why. As circumstance has forced me to look closer at what lies beneath my feet, so I have found different layers to unpeel. An interest that might once have stalled with a fascination of birds has deepened to encompass reptiles, butterflies, trees, fungi, flowers, geology and, ultimately, anthropology. And as I have stepped sideways into those folds, so I have found a greater freedom from myself and the constraints within which I often struggle. An interest in the moon, some broken pieces of calendar and an ancient people is more about me understanding about my place in the world.

We return to the canal path and head west, through more of the same that is also so completely different. The route is deceptive in its directness, perspective lost to the flat and the wide. A horizon that folds over itself like the edge of the sea. But it is glorious and solid with distraction. There is a threat of breaking cloud and sunshine but also, perhaps, of a rain shower. The former might just provoke a hobby into manoeuvre, but we seek the shelter of a hide to eat our lunch. Just in case.

This is a view I have not seen before, broken by trees and a slight rise of farmland with villages beyond. With a hatch open each, we have a shared direction of sight but with a decent scope from west to east. The north is well covered and should anything arrow as we eat sandwiches and crumble through a packet of stunningly good lemon and ginger biscuits, then it won't go unnoticed. And we don't wait long. A male marsh harrier, smaller than the female, with more grey and lacking the golden crown, drifts from the right. A steady quarter, then a pause, and in a moment a hover, pirouette and spiralled thump into the reeds. We gasp as one, but almost immediately it is airborne once more, a failed hit so back to the lollop.

A couple of buzzards drift high, drawing the attention of another harrier who wings up to gain some height advantage before folding its wings and firing a warning shot across their sterns. It is a fairly tame duel in truth, but were the buzzards any lower, and more of a threat, then the encounter might get a bit more feisty. Then, as we watch, comes the cut of a scimitar. Pointed wings swept back like crescent moons, short neck and a litheness of shoulder. The hobby has effortless speed. Smaller than the harriers and buzzards but 100 times more agile. Then a change in purpose, three then four deep pumps of the wings, before a stall as the talons shoot out and a target is hit. We all 'hear' the thump through our eyes even though the prey is too small for us to see. The hobby cruises as it quickly eats, before sharpening its eyes and sweeping back those wings once more. And it isn't alone. 'Woah!' Will cries. 'Where have they all come from?'

It's hobby hour and we are happy to indulge.

As I leave my friends, grins wide and memories cemented, I remember Mike again and the image of us leaning on the gate. A connection needn't have to come from a spectacular aerial display, although it's a pretty awesome thing to witness, but a small patch of green and something on which to lean will do. The cigarette isn't necessarily advised, but the friendship definitely is and Mike was an incredible friend in some of my darkest moments. He died last year. And I couldn't get to see him at his end. My brother, Rich, was there for him, though, to a wonderful degree. There had been a cruel twist around Christmas time as Mike found himself in intensive care, oblivious to all and oblivious to the fact that his best friend, another old friend of mine, lay in the adjacent ward. Mike did at least make it home, and I did at least manage to get to Jens's funeral before the lockdown that kept us all away from Mike's and kept so many from final farewells.

Two men too young. A reminder that moments such as today are worth absorbing every single blink of.

And still the cuckoo calls.

6

THE MOON OF HORSES

When I bestride him, I soar. I am a hawk: he trots the air; the earth sings when he touches it.

(William Shakespeare, *Henry V*)

I SWING THE DOORS open into another cloudless morning. The sun has not quite cleared the ridge, but the air is already warm and the oxeyes are patiently waiting. A huddled crowd of faces fixed in anticipation. I settle down beside them on the step, place my mug beside my feet, and then reach behind for my bowl of porridge and my binoculars. The former still steaming with a dollop of honey and sprinkle of cinnamon, the latter just in case.

There is a glisten of dew but it won't cling for long; today will be warm, very warm, and still. The sun fierce. And before I mouth a first spoonful of breakfast I pause to take in a decent noseful of the morning. It is actually quite difficult to describe. I can 'smell' the moisture, but not in the same way that a musty room or dank forest wafts. Nor the extraordinary tang of petrichor when long-awaited rain stirs the bacterial dust of aridity. This is clean, like the air above an upland stream, and works almost like a decongestant, clearing the passageways that then fill with the smell of daisies and nettles – all young and fresh, all vibrant in new growth. It is, essentially, a smell of life, and it is rather pleasant.

I rarely see this hour on a Sunday, not without the promise of pulling up the duvet and drifting back to sleep. But I'm feeling rather buoyed, and not just by the sun and warmth of early summer, but also the colours and sounds. This overlapping window, when the blackbird is still singing and blossom still floats from the trees, is a period I treasure each year. In a few days, on the 16th, the fishing season begins and my internal clock begins to operate at a different speed. Thoughts swirl with the water and, before I know it, the birds are quiet and haws, sloes and elderberries are leadening the trees. For now, though, I am without that distraction and I want to make the most of seeing the world without any particular purpose in mind. Beginning now with an unhurried cup of tea and should Sue also stir in the next little while then perhaps a second cup in her company. I blink skyward, the sun yet to break my horizon but edging closer, and scan around for a glimpse of the moon. It should have risen a short while ago but, in truth, I'm not entirely sure where to look. It could be obscured by the rise of the ridge or its thin silver sliver could easily hide in the haze of a mid-June morning. We are nearing the Celtic month of *Equos*, a word which no one seems to dispute the meaning of. What else could be derived from it? *Equus* being the Latin for 'horse' with close variatives in Irish Gaelic and Welsh, and derivatives still very much associated within the languages of today. It seems inconceivable that the engraving on the Coligny calendar could mean anything else, but what might it relate to?

The natural world was heavily deified by the Celts. Trees, for example, were regarded with huge importance by the druids who created an astrological system not unlike the modern zodiacal system (my 'birth tree' is the ash – hopefully the impact of dieback cannot be symbolically transferred). In Orkney, situated above the South Ronaldsay cliffs, is the 'Tomb of the Eagles', a Stone Age burial chamber discovered by local farmer Ronnie Simison, in 1958. Inside were around thirty human skulls, buried alongside the talons and bones of at least fourteen white-tailed eagles. Such ritual suggests that the eagles were placed in high regard, the purpose of their presence likely to protect or carry the souls of the dead into the afterlife.

Boars also carried important totemic properties. Small, carved figurines were fixed to helmets and armour, presumably as tokens of good fortune for battle. At Somme-Bionne in north-east France, the remains of a chieftain's burial tomb were uncovered and dated to around 500 BC. Inside, the body had been laid upon a chariot, with his weapons and horse trappings alongside him. The trappings were ornate in design, following a pattern of archaeological finds elsewhere in Europe. An unnecessary decoration unless the wearer was held in high regard. The horse, then, was clearly revered, and the skill of Celtic riders respected by the Romans who employed mercenaries from Gaul in order to exploit their skills. The Celtic goddess Epona was regarded as the protector of horses and she was also worshipped in Rome. A link that seemed to predate a natural hybridisation of respective cultures during occupation and perhaps suggesting instead that the Romans were influenced through a respect of the relationship that the Celts shared with their horses.

Still, though, the significance of a month so named *Equos* is somewhat puzzling. If the etymology was purely tributary then why pick this time of year? Links have been made between the lunar cycle and a foal's birth, with a belief that it coincides with a full moon. Perhaps, though, given that a mare gestates for around eleven months, this is the time of year when horses were encouraged to mate. That way, the foals would be born in May, beneath the Bright Moon, when the weather would be kind and food plentiful. Other theories point to a more symbolic meaning. *Equos* was a time of travel, a time to ride and explore. To meet old friends or re-establish trade routes or tribal pacts. It was a time to make the most of the long days preceding the busier late-summer months of harvest.

It could, of course, mean absolutely nothing, be a misunderstanding or an adaption from elsewhere. After all, we do still refer to the fourth day of the week being 'Thursday', the influence of Old Norse enduring in a culture that has never worshipped Thor, to whom the day was given. I rather like the idea of travel, though – the sense of freedom. It could sit in antithesis to the Quiet Moon, *Anagantios* – 'where one does not travel'. Now is the time to get out and about, a sentiment that seems particularly pertinent given the limitations that so many have faced over the last fifteen months. It pleases me in part, however, because it is the exact opposite of what I intend to do today. I will walk, but no further from the cottage than I could throw a stone. I have already explored the world around my doorstep, but today I want to share the grass beneath my feet – and there is much to see.

Some years ago I was invited to fish a rather lovely pond in the depths of Wiltshire. My friend Chris, who I met last month on the Avalon Marshes, had permission to fish there and that extended to an occasional guest. The journey slipped through ever-narrowing lanes until we creaked open an old gate and bumped along the edge of a couple of fields, the pool itself tucked among the trees and invisible to anyone who didn't know it was there. The fishing itself was superb, but the atmosphere of the place was what made it. We would creep into position and nudge to one side the reed stems to lower a float beside the lily pads, then inch back when finished and leave nothing but a patch of slightly flattened watermint. For several seasons I thought of nowhere else, even though I fished there rarely. Anywhere else I visited would pale in comparison. There would be other anglers for starters, the banks muddied and crumbling, the natural world a step further away. Engagement took longer and was invariably shallower, unlike the secret Wiltshire pool where spotted flycatchers pirouetted and water voles gnawed around your feet.

Then came a change of management and a keener interest in maintenance. It was fairly low key to begin with: a circular pathway cut back with a strimmer, the odd loose branch lopped and removed. But as is often humankind's habit, the controlling hands didn't know when to stop. Whole trees disappeared, areas of scrub and bramble razed to nothing. The fish stock of the lake was tinkered with, an imbalance forced. The pond was far from natural, originally formed by damming and then diverting a small stream, but the work that had been done since the creation of the place had been undertaken with sympathy to the habitat itself. It was, when I first visited with Chris, just about as perfect as it might ever be, in our eyes at least. Then, within a few years, it was transformed into a place neither of us wanted to visit. Hopefully the water voles found somewhere else to call home.

I have always had a low-maintenance ethos when it comes to land management, a stance that reflects personal preference as much as scientific opinion. It is not always beneficial, as I have learned while reading about hedge cutting and coppicing, but in the most part I feel that we do tend to over-manage. That fine line between conservation and control. Our little garden certainly reflects that ideology, though. The small rectangle of lawn is mown regularly, but the remainder is left to do as it wishes for much of the year. I will clear the patio each winter, waiting until the vegetation has died back to a point of ease on my part, but at this time of year it is a riot of loveliness and we never know quite what will pop up. The slabs are laid on to the earth with plenty of gaps for self-seeding plants, many of which are feral. There is a glut of wild strawberries, fennel, cloves, borage and marjoram. In the late summer and early autumn, the fennel is especially attractive to chiffchaffs who flit from umbel to umbel, taking food from the flowers and seed heads themselves, as well as the insects that are the more obvious attraction.

For the time being, though, the patio belongs to the oxeye daisies. As I sit among them so they seem to smile in response to the sun. This is a perfect platform for them, slightly raised, facing south by south-east and with the heat of the sun in the south bouncing back from the stonework of the cottage. If I were a daisy, I'd happily make my roots here. The fresh shoots have also provided food for a couple of very welcome guests this spring. Two rabbit kits appeared on the patio a couple of months ago and took refuge beneath the rather decrepit old dog kennel that doubles as a lawn-mower shed. They were tiny, obviously fresh from the warren, and improbably cute. And the spread of greens on the patio table provided sufficient food for them to eat in apparent safety. As the days warmed and we began to open the French doors they continued to appear, content to be within a few feet providing we were quiet and our movements unhurried and deliberate. A sad day came when returning from the supermarket and spotting one of them drowned in the little ornamental pond that sits close to our back gate. Whether it had been chased and jumped the low hedge that surrounds the water, or had simply explored where it oughtn't, I couldn't tell, but it cut a sad, bedraggled image as it lay floating and tangled in the pond weed. Further upset came as I stepped over to reach out its body and lowered my foot directly on to a sunbathing grass snake. I managed to limit the downward pressure and the snake launched itself into the water, a head gingerly reappearing at the far end, tongue flicking fast. She was the snake that had hibernated in our garden, possibly down an old rat hole, as we had seen her basking in the late winter until she took up residence in the little pond. She is distinctive in colour, a light, lime green and around 40 inches in length. I was horrified at the possibility that I had hurt her, particularly as I had found her mating a fortnight or so earlier. I was, fortunately, wearing only my flip-flops so felt the resistance immediately and when I snuck back a few hours later she was basking once more seemingly untroubled and with no obvious signs of injury.

The remaining rabbit (we named it 'Stewie') has remained, although it is already too big to squeeze beneath the kennel. Instead, it has extended its range to include our landlord's chicken coop that sits adjacent to our garden, finding more comfortable headroom beneath one of the chicken houses. It regularly nibbles on the patio still, remaining tolerant but with increasing twitchiness, and probably wouldn't have caused the excitement it has without the dramatic downturn in rabbit fortunes over the last couple of decades. Research from the British Trust for Ornithology has found a reduction in the rabbit population of over 60 per cent since the mid-1990s. Myxomatosis, long associated with dramatic swings in rabbit fortunes, is likely a factor, alongside rabbit haemorrhagic disease virus, which has become increasingly prevalent. But, as is often the case with more abundant species, we tend not to notice the decline until it becomes too difficult to determine the drivers behind it. The rabbit population in the village had disappeared quite quickly (we have only been here for a decade, after all) so their lack was more conspicuous, and it is very lovely to have them back, although their presence is not universally welcomed. Their fondness for the most green and succulent shoots is often to the dismay of gardeners and farmers, and their numbers are often controlled as a result. On the Isle of Portland, meanwhile, only 20 miles or so from where I am sitting, the rabbit is considered something of an ill omen. It is suggested that the presence of rabbits in the great stone quarries for which Portland is renowned could weaken the ground and cause danger to the quarrymen, resulting in the cry of 'Rabbits!' to be portentous. As a result, Portland islanders refer to them as 'furry things' or 'underground mutton', a custom not lost on the promoters of the Aardman animation *The Curse of the Were-Rabbit*, who sought alternative phrasing for the publicity of the 2005 film on the island. Perhaps, given the population of Portland is fewer than 13,000, the resulting publicity did rather better for the nationwide cause rather than influence those cinema goers on a small Dorset off-shoot.

For an animal so ingrained within our contemporary culture, it is odd to consider that the rabbit is not native to the British Isles. There has been much speculation as to when they arrived, but tests carried out in 2019 on a piece of tibia found at Fishbourne Roman Palace in West Sussex determined that it came from an animal alive in the first century AD. It was long presumed that the rabbit had been introduced by the Normans rather than the Romans, as there was no record prior to the eleventh century and nothing recorded in the Domesday Book of 1086. They became an animal popular for their meat and fur, and warrens were created in the Middle Ages for their keep, with the term familiar today in the names of farms and villages. The seaside resort of Dawlish Warren, in Devon, is a familiar example, while Black Rabbit Warren in Norfolk points perhaps to the importance once placed on the value of different-coloured pelts – the melanism that would cause the black colouration being genetic and therefore able to be isolated and that strain specifically bred.

Sue and I have tried hard not to get too attached to Stewie the rabbit, but that has already failed with the naming. It's a dangerous business attaching sentiment to a wild animal, but it is also slightly inevitable. There is a privilege to be enjoyed when other living things share our space. The rabbit on the patio, the sparrows on the bird table, the blackbirds that raised three broods in the hedge last year. The rats that sometimes clatter around in the loft are less welcome, but we have grown attached to the house spider that lives behind the toilet cistern, and happily share the bathroom with a score of cellar spiders. Silent assassins who sit unmoved for weeks and months and then spin a cocoon around a woodlouse or crane fly in a matter of seconds. Naming them adds another level of attachment, though, which inevitably ends in upset. And in the case of Stewie, it is also somewhat inaccurate, because as he has grown, so it appears that he is actually a she.

One of the interesting facets of recent debates on the presentation of history, and more specifically the whitewashing of history, is that of discovery. Rather like the nine planets that *don't* actually make up our solar system, certain misapprehensions apply in regard to Western exploration. Columbus and North America, the Conquistadors in the South, or Cook in Australia. All 'discoverers' of places that were already inhabited by humankind, and all names and associations that are etched into the memory of most people of a certain age. This arrogance of Western culture actually extends further. In order to reason as to why civilisations within these places were seemingly more advanced than our own may have been, by constructing great pyramids in the desert or citadels among the Andean peaks, we develop elaborate otherworldly explanations. And without archaeological evidence, or by taking the Roman scribes at their word, we might easily overlook the complexities of Celtic culture – particularly in regard to travel. If *Equos* was a time to move about, to venture beyond the boundaries of normal, then the people who lived here, or on Eggardon at least, would have likely pushed a boat out on to the sea at Cogden as walked or ridden up to the Avalon Marshes.

The Atlantic Bronze Age system, as described by Barry Cunliffe, covered the British Isles, what is now northern and western France, and the western part of the Iberian Peninsula. The common cultures found through this area point to extensive trade and travel, much of which would have had to have been made across the sea. A key determinate of the Celtic culture might be found upon their coins, and finds within the realm of the Durotriges carry images of horses and boats. They are stylised and on occasion, if I am honest, beyond my unguided comprehension, but the designs are both beautiful and complex. Rather like the horse trappings and intricate weapon gilding discovered within places of burial, the world of the Celts, though at times brutal, was far more advanced than we might believe from what is written.

I do like that thought. Not that it has any great bearing on my life other than helping me understand the way the world works today. Pondering the existence of a time that was gives me perspective of the time that is. And the world at present is certainly unsettling. Not that I'd fancy a timeslip into the past. I'm no fighter, no great hunter (fishing aside) and I'm not sure mental illness would be empathised among the blood and guts of Celtic life. Perhaps the druids might have brewed up a potion for me, but I would have likely ended up an outcast or trussed up on a sacrificial altar. Either of which might be preferable to venturing out to the coast on this sunny Sunday as the Durotriges may have done 2,000 years ago.

We knew when we moved here that we would be sharing our new-found space in the holiday months, and Sue spent her formative years in Torbay so was well used to an annual influx of strangers. But the Dorset coast has been especially busy during the past fifteen months. To people locked down in homes and high rises, holed up without gardens, without space to breathe, the south coast must seem utopian. With that in mind, we cannot begrudge the hordes, and the local seashore seems to have become *the* place to head. The litter is unwelcome, however, as is the disregard for hygiene. With public toilets having been closed, the beaches and edgelands have become a literal dumping ground. Sue and I ventured to the sea, as always, for Sue's birthday ten days or so since. The following day we both had poorly tummies and realised it was likely a result of sitting ourselves on the pebbles as we always do. We hadn't considered the fact that the makeshift tented towns, which have popped up last year and this, are without sanitation.

I've walked around 50 yards from our back gate, across the grass that surrounds the ornamental pond and to a gap in the hedge that flanks the eastern edge. The grass remains damp in the shadows, the blades clinging to the moisture but ceding plenty to my flip-flopped feet. The sun had caught me in contemplation, sneaking over the brow of the ridge before I had crept over to check on the snakes. The first few minutes of direct sunlight are key to grass-snake spotting at this time of year. On the compost heap here at least. This one has been mulching the estate waste for far more years than we have been living here and it provides the perfect place for egg incubation among a wider habitat that is ideal for grass snakes. Beneath the slope that drops away from me are a couple of lakes, formed half a century or so ago by damming the spring that bubbles up and supplies much of the village with water. I have fished them, of course, but the novelty of angling on our doorstep didn't last. I got to know the lakes and the life within on a different level, a more personal immersion. Without the separation of travel, the way in which a journey can help create purpose, I had no urge to catch the fish. Instead, I would creep the margins and watch the patterns of the seasons unfold. The bluebell carpet on the northern bank, the mare's tail on the south bank that bursts the soil like a spear and then opens like a spindly Christmas tree. There is the first comma of the year soaking up the sun on the bare earth of the dam and the piped call of a kingfisher flushed and flashing blue as it needles to a new perch. During our first year here, while I was still fishing the pools, I kept a keen eye out for the grass snakes. I had been told that there were one or two about and caught several glimpses, usually a movement in the margin or sun-drenched bank, but occasionally a better view of the sweep across the surface, body swaying and head held just above the surface. There was an absolute mass of available food. Aside from the carp, eels and rudd, the water would be thick with amphibians. A mass of frogs, toads and smooth newts, the perfect menu for a grass snake – easy to grab and then digest. What I hadn't appreciated, though, was that several of the females would make their way up to the compost heap behind the hedge each spring to lay their eggs, drawing suitors as they did so. By edging into a slight gap in the hedge (possibly widened by my edging) and keeping still, I could watch as the snakes slithered in and out of the heap. The first couple of hours of morning would be best, before the sun worked around to the west and by midday left the heap in shadow. I was surprised at how many snakes there were, the subtle differences of colour and shape alluding to that fact, even if I rarely saw many individuals at once. On more than one occasion I had a snake slither across my bare toes (I remain in flip-flops, often stubbornly, until well into autumn), which is a curious sensation particularly when unexpected.

Finding grass snakes under my feet – as well as on top of them – was a revelation. They were a species I had often encountered, but invariably within moments that were fleeting. Now, though, I had the opportunity to study them. To identify them as individuals and perhaps understand them a little better. It is likely that the bigger females, some of which might be 15 to 20 years old, had been using this compost heap to nest in for their whole adult lives, with generations of their offspring then returning to the spot where they first hatched each year. Emerging in complete independence yet driven and drawn by instinct and the influence of cold blood.

Between me and that compost heap, running along the line of the hedge, is a raised bank comprised of excess soil that was salvaged from a local house excavation. Ashley, who owns this estate, had no initial plans for it but, having seen the extraordinary bloom that appeared the following spring, left it be as a thing of floral colonisation. There was originally an explosion of common poppies, their seeds having been stirred from the still of interment. They may have been buried deep for decades before the disturbance gave them opportunity to germinate. Other pioneers were windblown – thistles and ragwort – the cotton-fluffed seed heads having floated into place soon after the bank was set. Each year has seen a shift of species, with grasses making a steady smother, but the raised nature and chalk body offer a perfect basking spot for grass snakes. Not that they are a reptile that often courts direct sunlight; away from the early rays (which I have likely missed), they will tend to favour partial or complete cover. Not nearly so confident in their camouflage as an adder might be. On the end of the bank are a few small scraps of old roofing felt, placed with this in mind. The pieces are small, none bigger than a dinner plate, but beneath one will be at least one grass snake, albeit only tiny itself, and as I lift the edge of the felt to peer underneath there are actually two, coiled together with a similarly sized slow worm for company. All three reptiles have emerged this spring from their first hibernation, the slow worm – a thin sliver of bronze-lined gold – likely older than the snakelings but only by a few weeks. I found three hatchling grass snakes beneath this felt last autumn and all three made it through the winter. They should disperse any time soon, perhaps the third one has already, and I will be glad to see them gone. They seem quite vulnerable in this spot, and a careless footfall from human, dog, deer or even rabbit might squash them. Although in the open, when they do move off, they will face even more danger. Aside from the more obvious predators are pheasants, rats and even blackbirds, all happy to make a meal of a 6-inch snake. With this in mind, I lay the felt back upon them before they stir, a ploy I will not undertake in a few moments as I move into the old goose pen.

Here is a lovely triangle no bigger than a tennis court, which Ashley has also let be. Since the geese were moved out seven or eight years ago, the ants have built their hills and the grasses are dotted with colour. For now the white of cow parsley has given way to the rouge of red campion, but in another month or so will come the blues and mauves of knapweed and scabious – just in time to feed the emergence of marbled whites. My path through the grass is well trodden, leading from the gate to a pair of tins that were laid a couple of years ago for the use of the grass snakes.

They get very well used, but this morning I have already left it too late. The sun has warmed the metal to a point beyond comfort, even for the cold-blooded. A single slow worm lies beneath one of the two, but the only trace of snake is the pair of skins that are stretched and torn through the tangle of grass. I wait for the slow worm to move away before lowering the tin; the undulations of the corrugated metal might easily pin the reptile down as I replace it. A fate that has befallen two snakes this spring, although the downward pressure was likely caused by the slots of a roe deer. One of the snakes survived but with a distinctive kink in its neck, but has since sloughed and, presumably, eaten. A smaller snake, fresh from a second hibernation, was not so fortunate and likely perished from the cold. Just as a piece of tin soaks up the rays of the sun so it cools to ice in response to dew and frost. I brush back through the scentless blush of red campion, an appropriate flower given the ground on which it grows. It has long had an association with snakes, particularly in the south-west where the seeds would be crushed to create an anti-venom for adder bites, a misfortune believed more likely should a flower be picked and taken indoors. Meanwhile, on the Isle of Man the red campion is traditionally known as *blaa ny ferrishyn* – 'fairy's flower' – for the flowers were where fairies would keep safe their stash of honey.

The kettle is on and the coffee pot primed. Mid-morning and the day is already warm, the south-facing stone walls of our cottage trapping the heat and offering a waft of the Mediterranean, the sun chasing the lightest of breezes from beyond the sea. These are the days that seem so distant in the colder months. When we huddle in hats and big jumpers, an extra pair of socks and muddied wellies on the mat. Today the doors will stay open until bedtime, the whole house breathing in that soporific air.

Coffee smells good in the sunshine, mixing with the flush of fresh nettle growth and a hint of fennel. A buzzard mews from the blue, too high for my eyes to pick out, and the courtyard swallows are busy in chatter. It is the sort of day when a hobby might rip through, and the swallows will soon alert us if one does. They are quick to challenge any passing raptor but respond quite differently to each species. Buzzards will usually get a polite escort, their threat seemingly deemed moderate, whereas sparrowhawks are harried and hassled until they are some distance off their territory. Kestrels and peregrines will be mobbed but not with the ferocity of the sparrowhawk, whereas the hobby elicits a reaction of deep respect. It is the one species that the swallows will be familiar with throughout the year and also the falcon most likely to be an aerial threat. As a result, the swallows and house martins look to gain the advantage of altitude, staying above the hobby and diving down on it rather than risking an all-out chase that might surrender airspace.

A movement on the patio catches my attention, although it is as much the sound that alerts me. A distinctive scratchy crackle of a body slithering. I have to smile; having gone looking for a snake (admittedly only paces away), one has now come to me. I stay still and watch as the grass snake moves slowly across the slabs before slipping into the thick of goldenrod that dominates the 'flower bed' and out of sight. It was a male by the looks of it, slim in body and long in tail, and probably not much more than 24 inches in length. I would guess that he was the individual that mated with the female beside the ornamental pond, though quite why he is on the move this morning is uncertain. Perhaps he was disturbed or found the spot where he had been skulking getting rather too hot for comfort. The cool, damp earth where he has slithered will be far more agreeable on a hot day such as this.

A couple of years ago, having had the doors open on a day like today, a grass snake had, unbeknown to us, made its way inside and into the lounge. The following morning, with the doors closed to less pleasant weather, it slithered out from behind the television as Sue ate her breakfast on the sofa. Sue didn't panic but called me in to usher it outside, by which time the snake had coiled behind the other settee, peering up at me with an expression of disgruntled interruption. We felt rather privileged by the visit, although there are many people who would not. Snakes do elicit a strong unconscious response, often of fright or flight, and with valid reason. Almost one-fifth of the world's snake species are venomous and according to the World Health Organization between 81,000 and 138,000 people die around the world each year as a result of snake bites, with many more facing amputations or long-term health issues. For the most part, these incidents are the result of accidental encounters that cause the snake to bite in self-defence, with greater fatalities occurring in remote, rural locations with poor access to health facilities and vital anti-venoms. It is unsurprising then that the snake is so often portrayed in a poor light, from tempting Adam and Eve with an apple in the Garden of Eden to the giant snake Nagini in the *Harry Potter* books. Yet this is not always the case. The Celts held the snake in great esteem and one belief points to the universe being born from two crimson eggs laid by a serpent in the crook of a willow tree. The first egg hatched to release the sun while the second bore the earth. The Celtic symbol *Wuivre* comprises two intertwined snakes, one dark, one pale, and is believed to signify the qualities of the earth, while the druids would often carry an 'adder stone', also known as a hag or witch stone. These were stones with natural holes through the middle which were believed to hold incredible power. It was thought that these stones were formed during the balling of snakes (presumably adders) and that special measures had to occur by which the stone or egg might be gathered. Pliny wrote of the process:

> A vast number of serpents are twisted together in summer, and coiled up in an artificial knot by their saliva and slime; and this is called 'the serpent's egg'. The druids say that it is tossed in the air with hissings and must be caught in a cloak before it touches the earth. The person who thus intercepts it, flies on horseback; for the serpents will pursue him until prevented by intervening water. This egg, though bound in gold will swim against the stream. And the magi are cunning to conceal their frauds, they give out that this egg must be

> obtained at a certain age of the moon. I have seen that egg as large and as round as a common sized apple, in a chequered cartilaginous cover, and worn by the Druids.

Such an unlikely course of events would be beyond the abilities, and common sense, of mortal beings of course. After all, no one in their right mind is likely to grab a ball of writhing snakes, throw it skyward and then attempt to catch it again. Ensuring, therefore, that the mythical nature of an adder stone endured.

Snakes were also revered by the Romans, and it is possible that a fourth species of snake was brought here by them to slither alongside the native adder, grass snake and smooth snake.

The Aesculapian snake is found across much of mainland Europe, including France, southern Germany, northern and central Italy, and through the Balkans, and its distribution can be largely attributed to the Romans.

A plague had decimated part of the Roman Empire, and in desperation they sent a ship to Greece in order to seek the aid of Aesculapius, the god of healing.

Aesculapius came on to their boat in the form of a snake and returned with them to Rome where he took up residence on an island in the Tiber estuary, the plague subsequently abating. To thank him for his help, the Romans built a temple in his honour, and subsequently took Aesculapian snakes in earthenware pots to temples and baths where they were released in the belief that they were representatives of Aesculapius himself, and therefore carried his powers of healing. They also kept rodent numbers down, the rat being their most favoured food, and this in turn would have checked the spread of disease and justified the belief in the snake's healing quality.

The rod of Aesculapius, a staff with a snake entwined around the shaft, became a symbol of healing and medicine, and is still used by organisations around the world today, most familiarly on the Star of Life, which can be seen on the ambulances of the NHS. Today, there are two feral colonies of Aesculapian snakes in the

British Isles. One beside the Welsh Mountain Zoo at Colwyn Bay in North Wales (the origin of which seems fairly apparent) and a second in London, with a well-established population living alongside the Grand Union Canal in Camden and Regent's Park. There is greater uncertainty surrounding the origin of the London snakes, although one of the best sites to spot one is within the grounds of London Zoo. An escape from there seems rather more likely than a survival since the Roman occupation.

It is amazing how easily a lazy Sunday morning disappears into a dreamy Sunday afternoon, my efforts to counter the notion of *Equos* adventure having become even more acute. I'm not even likely to make it back out of the gate to study today's flight of butterflies and in many ways I have no need to. As I sat and drank my coffee I watched the soft flutter of two cinnabar moths working the patio and showing me which of the green sprouts will later bear the yellow flowers of ragwort. The flight of a moth seems even softer in daylight, the scarlet hind wings as delicate as silk. I must have spent half an hour watching them move from plant to plant, laying tiny dots beneath the leaves, while the smiling faces of the oxeyes sway softly around them.

Sometimes, there is no need to go anywhere.

7

THE MOON OF CLAIMING

Nature alone is antique, and the oldest art a mushroom.

(Thomas Carlyle, *Sartor Resartus*)

I AM STANDING beside a vast swathe of beeches and if I close my eyes then I could imagine myself beside the sea. The salty edge is lacking, while the breeze is too light and the air too heavy to have brushed onshore, but there is a tenuous aural connection to stir with the sun on my skin. The rattle of pebbles beneath the surf is coming from not a beach or the beeches, but directly above me, from the leaves of a single aspen. It is an unassuming tree, quite tall for the species but slim-trunked and sparsely branched. The bark is unremarkable, marked here and there but generally smooth and grey, pointing to a lack of weathering that comes when you are short-lived. And I might in all honesty have walked straight past but for the sound of those leaves. Themselves small and rounded, with a lightly serrated edge and long, slender stalks that flatten at right angles to the blade. A feature that aids the leaf's movement, for though they might be singularly non-descript, as one they flutter in response to the softest breath. A lone clatter in an otherwise silent forest, and as I pause to listen, I understand why the Celts may have believed that the aspen leaves rattle as they communicate with the next world. There is movement even when the air seems stopped still. *Populus tremula* indeed.

It is a beautiful summer's day. A clear sky, a few wisps of white tone down the blue, but the sun, though hot, is tempered by the direction of that breeze. The lightest of northerlies that feels so clean against the stifle from the south. Not that it is cool enough for comfortable walking, though. On a day such as today there are only two realistic choices. The coolness of altitude, up on to Eggardon where I would be sharing the air with others, or into the shade of the trees, where I will tread a path alone. I take a couple of steps through the squelch of a ditch and move happily beneath the peace of the beeches.

It is estimated that there are 3.2 million hectares of woodland in Britain, slightly less than half of which might be classed as native woodland. The total coverage represents around 12 per cent of the British landscape which compares to a peak 6,000 years ago that was likely as much as 75 per cent. Aside from the significant difference in area lost in that intervening period is the fragmentation. There is little in the way of extensive, mature forest remaining and instead are isolated pockets that cannot sustain the same ecological depth. Trees are long-lived, with species such as oak and yew exceptionally so; as a result, natural woodland cannot be simply replaced. That is not to say that we shouldn't plant trees, but we should value existing habitats more than we seem to do. There is, for example, a huge area to the south-east of our cottage, the other side of the Roman road, that has been planted with trees as an area of conservation. It is somewhere in the region of 500 hectares in size (a rather rough estimate using an Ordnance Survey map and a piece of cotton) and is a wonderful project to undertake. It is sobering, though, to consider that the benefits of the creation, in terms of ecology, might not be fully appreciated for a century or more. I have walked through the older, more established parts and there is plenty to see and hear. The leaves of the trees offer food for caterpillars and aphids, drawing larger predatory insects that themselves are eaten by small birds and mammals. The coverage feels a little superficial, though, and in winter it feels sparse – the bare trees standing skeletal upon a floor that is waiting patiently for an identity. As the leaves fall each autumn they will draw windborn fungal spores that will bring about mycorrhizal process, a symbiotic association of fungi and plants that can continue indefinitely. What is missing for now, though, is an understorey. Those plant species that grow beneath the woodland canopy where little light will penetrate. Mosses, ferns and vines, flowers such as lesser celandine and wood anemone, and smaller tree species like holly or dogwood. As these species colonise so they help maintain the humidity, encouraging nutrient recycling that is welcomed by all floral life. With more life comes the process of death, invertebrates that feed upon decaying matter and again provide food further up the chain. For this to happen trees need space and too often we plant without a bigger focus in mind. Rather like seeding barley or wheat, squeezing as many saplings into as small a space as possible. Ticking boxes in terms of number but not considering the ecological need. Something is always better than nothing but planting trees in the right places, and not at the expense of existing habitat, is vital. And, of course, allowing existing woodland to endure is more worthy than anything else.

I lean back against a beech trunk. It is reassuringly solid and the bark refreshingly cool. A beech tree might live for 250 years or more, but despite this one's size, it is likely to be less than 70 years old. I know this because this area of woodland was almost completely cleared at the end of the Second World War, in response to a national shortage of building materials and an unprecedented need for rebuilding. Fortunately, however, systemic replanting took place soon afterwards, and although made with further harvesting in mind, this did ensure that the habitat remained. The raw ecological materials were still in place and, though the trees might not be that old, the 'forest' itself is ancient. And it certainly feels it. Ten yards or so in front of me is a steep bank that drops into a narrow, but deep, ditch. There is a trickle of water that is presumably spring-fed and has, over many years, bored a deep cleft into the greensand substrate. There are several similar watercourses cutting through the forest, all of which provide excellent drainage. The ground here will get wet after rain, but only in a few places will water sit for any period. So it is that the waste wood, the fallen branches and storm debris, will break down gradually rather than be rotted by immersion. Both beech and birch, which are the dominant deciduous trees, are prone to rot, their broken structures sucking up water like sponges. Dry air can slow the process, though, and offer greater opportunity for other life to make use of the fallen. And there is far more going on around my feet than high above in the green of the canopy. For a moment or two, though, I will pay heed to the tree that props me.

The beech is fond of chalk or lime, suggesting that the reddened clay beneath my boots is topping another substrate like coffee icing on a sponge cake. And that icing is capped by the crunch of spent masts and fallen leaves. The soil beneath a mature beech is kept rich by the tree itself, which offers shade to the edge of the dripline and that thick blanket of leaves to break down into matter. As a result, the beech has long been known as the 'Mother of the Forest', and the nineteenth-century German forester and professor Karl Gayer spoke highly of it:

> Without beech there can no more be properly tended forests of broad-leaved genera, as along with it would have to be given up many other valuable timber-trees, whose production is only possible with the aid of beech.

Gayer studied among the great forests of Bavaria and exerted great influence on forest management. He focussed upon the biology of the forest as a whole, considering the relationship between species and the importance of soil and mixed woodland. He also believed that natural rejuvenation was the best method by which to make a forest productive, a pattern that has likely, in part, led to the make-up of my surroundings. There are large swathes of pine and spruce, grown presumably for timber production and growing more often in the lower, damper dips, but by and large there is a sense of self-sustenance. And although the beeches are well spaced, their branches and leaves stretch to form a dense cover shading the ground beneath. As a result there are few signs of fresh beech growth, but other understorey species have been more successful, with holly particularly prevalent. There are several small trees within a few paces, although none are any taller than me. The holly's small size enables it to find a niche here, although where it finds the room to stretch for itself it might grow to 40 or even 50 feet. Interestingly, the spines that edge the deep emerald leaves and, alongside the red berries, make the holly so distinctive, disappear with height. Once a tree reaches around 10 feet, the leaves grow smooth, as though, having reached a point where the mouths of cattle or deer cannot reach, the holly knows it can economise its leaf production.

There is a sense of impenetrability to a holly. The stiff branches and dense leaf growth ensure that those spines are almost impossible to negotiate without being pricked. Perhaps this inviolability is one reason why the holly was so revered by the Celts and the druids believed it offered protection from malevolent spirits. Wreaths were formed and worn by chieftains to provide good fortune, similarly practised in Roman and Greek cultures and enduring to this day as a festive decoration, though hung on doors rather than worn on the crown.

It isn't alone in the understorey, though. There are several ferns, particularly along the banks of the ditch, and a coverage of moss that in places looks deep enough to lie down upon and sleep. Brambles tangle but do not have a strong foothold, though there is a sedge species that seems to dominate in patches. I have encountered it elsewhere, at Powerstock Common and also in our garden, where it is a relentless grower. It seems to sprout unnoticed before sending out seed heads, on long, stiff arms, that spill into every nook and crevice. In these woods, it does at least seem limited to the damper spots, and elsewhere is found a more even spread of life and, just as importantly, death. Those dropped limbs and storm-flattened trunks are left to the process of time, and as the days shorten and the air dampens, so they will be adorned with the thousands of fruiting bodies that the fungi living within the rot produce. And despite the time of year, and the heat, one reason I am walking here today is for the mushrooms. I might not be much of a gardener, but I do have an urge to tend the crop here as though it were my own.

Were we to be living today in a landscape dominated by trees, then perhaps we might feel a deeper connection to them. As it is, trees are often decorative or inhibitive, blocking views or causing neighbourly disputes. We might make use of the timber, use the limbs for fuel or be felling to prevent the further spread of disease, but so do we chop and clear on a whim. A life that might be 1,000 years old ended in a minute by a capricious chainsaw and no pause for thought.

It is, of course, impossible to compare then and now as like for like. The population of Roman Britain is estimated to have been around 4 million people, shrinking by the time of the Domesday Book and increasing only to 10.9 million by the time of the first national census in 1801. Today there are more than 67 million people living in the United Kingdom and such density, alongside the amount of agricultural land required to feed them, does not leave much room for wild space. In fact, the concept of 'wild' is something of a paradox. Today, almost as much land in the UK is used for agriculture as was forested 6,000 years ago, so a true 'wild' view is seen by few. Certainly, here in Dorset, despite the lack of intensive land use, I will struggle to find land that is not utilised for some purpose, and less still will I find 'common' land where I might walk and act without trespass. An individual's right to roam is an oft-discussed and divisive argument, and far too complicated to unpick here, but the freedoms that we would once have enjoyed would come with some heavy caveats. There might have come the risk of ending up as something else's dinner, while a slip or fall could have had terminal consequences.

I love the idea of existing among such an old landscape, though. Of feeling as though I were trespassing not because of private ownership but because no feet had trodden there before. I once stepped into the trees of a deep gorge in the Wye Valley – part of the Forest of Dean. The land there was truly ancient and the atmosphere quite overwhelming. I had been fishing, so my mind was perhaps waterlogged and not fully prepared, but still I lacked the receptivity. This was a place where people simply didn't tread, and as a person, I felt unwelcome. A tarnish. I rather liked it and I bowed an apology and left.

An understanding of the trees, and the life within and upon them, would have been vital to the Celts. Although farming and animal husbandry were core to their existence, supplementing a diet from the natural larder would have been essential. There were no pesticides or chemical fertilisers, no weather modelling or on-call vets. Blight among crops or disease in livestock could be catastrophic, but indiscriminately picking berries from shrubs or mushrooms from the forest floor was not a viable alternative. A mistake might only be made once. Knowledge would, therefore, have been precious, and presumably shared beyond the inner druidical circles. The identification, and naming, of different species would have had to go deeper than the commonly encountered. My own knowledge of fungi, developed almost wholly for culinary gain, would have been fairly basic two millennia ago and I do not possess any of the extended skills of many foragers in regard to plants, roots, saps and tubers.

I did, in fairness, take that initial interest in mushrooms as a means of extending our self-sufficiency. With Sue's income gone and her being the one of us with a career and greater opportunity, we found ourselves in the low-earning trap that many are stuck in. The media spin that depicts people sponging off the state only tells a small part of the story. Yes, there are some who milk the system, but they are few. Far more are the people without a trade or qualifications to ever earn a comfortable income. The cost of living increases within another's employ – childcare, journeys to and from workplace, clothing, food at work and more convenient food when at home. With a spouse or partner ill at home there is no saving on daytime energy costs, there are more meals that need to be pre-prepared and there is that dreadful stigma that attaches to genuine hardship. We do have it better than most in the world, yet we live in a society that treats illness as a weakness. Something to be shamed or scorned. State assistance is ever more temporary, with the premise of 'helping people back to work' but with a strict time limit attached. Fail to fit into a curiously designed model and the system will spit you out – it can't have poorly people boosting unemployment figures. The absurd reality in this country is that millions of people are stuck. The gap between the place they are and the place that society, or they themselves, might point them is far too great. A new opportunity might require full-time childcare or a second car, things that cost far more than the opportunity can provide. There is an unfortunate truth to the phrase 'I'd love to work but I cannot afford to', yet it is scorned from a place of secure privilege.

It must be said, though, that scrimping can be incredibly satisfying – and I've become quite useful with a needle and thread and a pair of scissors. It also helps us to appreciate things, by not just value but time. And combining the two, saving money and having fun at the same time, is the greatest prize. Which is why I began searching the woods for mushrooms – suddenly a lunchtime soup had flavour and a bowl of rice became a wild-mushroom risotto. Empty jars and Tupperwares would be crammed with dried delicacies and we could give gifts again at Christmas.

Woodland mushroom species tend to pop up in specific spots. The fungal mycelium, the network of strands that lives within the soil, spreads itself wherever able, but throws up its fruiting bodies – the mushrooms – at points where it is hindered. Banks, footpaths or sunken ditches can check tendril spread and so are perfect places to look. Factor in the soil type and tree species and you begin to see a woodland habitat very differently. And these woods, dominated by the beech, are perfect for one of the finest eating mushrooms of all – *Cantharellus cibarius*, sometimes known as the girolle, but more commonly called the chanterelle. They were a species I longed to find and as I explored the local area I often would. Ones and twos, here and there. They are egg-yolk yellow in colour and look striking among the green of moss or the brown of fallen leaves. Thick stalks with forked, false gills that are folds of the actual body rather than separate, detachable parts. A large chanterelle opens out to almost demand attention and they are impossible for a mushroom addict to ignore. I actually found a decent patch on an early foray close to home but was so excited by my find that I forgot to get a proper sense of where I was. I knew it was in a slight hollow beside a holly bush close to birch and oak, but when I returned I found lots of hollows and they all seemed to have holly for company. I went back several times and systematically searched the area I thought I had been in but could only deduce that my bearings had hopelessly failed or I had stumbled upon a parallel fungi universe. There was a chance, of course, that it was someone else's 'patch' subsequently picked, but I doubt that. Aside from the remote isolation was the general code of foraging conduct. There is no need to pick everything. Leave some for others to find and leave even more so that the spores disperse and the mushroom fulfils its primary purpose and provides food for the slugs and woodlice.

My hunt for another golden patch took me deeper into the local and not-so-local countryside over half a dozen consecutive seasons. I found odd chanterelles, but never any great number, until I wandered beneath these trees. There were only a few to be found on my first visit, but it was late in the autumn and I guessed that I had missed the main flush. The following year I visited a month or so earlier and was suitably rewarded, but it wasn't until June of the next year, when I came to simply enjoy the trees, that I realised just how early they began to appear. The first few nudge above the leaf litter in late May, almost at the moment that the beech-leaf canopy has fully thickened. By nurturing these, treating them rather like a vegetable plot, I can thin out the dense clumps that hinder one another and allow individuals to grow bigger and bigger. In doing so, I am of course taking the weekly excess home for the kitchen and, providing the weather does not get too extreme, I might still be visiting in early November. For five months we have a steady supply of gourmet mushrooms and get a bit blasé about how we eat them. They do need moisture, though, and this weather is set for some time with temperatures rising higher still. I'm going to pick a few more than I usually might at this stage of the season because they might end up desiccated in another week. And I'm going to cover a few patches with the remnants of last autumn's leaf fall, just to give them a bit more of a chance.

There is satisfaction to be found in tending this crop but I miss the magic of discovery. Of walking through trees that are strangers to me, not sure what might be lurking beneath. Looking for specific tree species and the mushrooms associated with them, and those features that might have prompted a mycorrhizal response. It is very similar to angling, where fish, being cold-blooded, can be far more predictable in habit and location than mammalian life. Rather than learning to read the river, though, as I have spent several decades doing, I was learning to read a woodland. It was mildly intoxicating and for a few autumns I thought little of water. The problem comes, however, from reading the same story so often that you cut straight to the ending. I know precisely when and where I can fill a basket so now I have stopped searching and all but stopped reading.

My route through these woods is familiar. A circuitous path that links each mushroom spot with optimum efficiency, but leaving the best late-season spot until last. That way, when the first bites of winter do see off the fungi, I should still have a treat at the end of the walk. I follow the top of the bank where I enter the trees up and deeper into the beeches, the slither of water in the ditch some 15 feet below me. I pause briefly beside a tree stump, the result of windfall rather than forester hand, the sharded edges and uneven surface no work of a saw. A small lump of the trunk sits beside it, a form long since weathered and rotted into misshape. Like the last blob of a snow drift, dirtied but defiant, a slow thaw into nothing. The cleanest edge of the stump thickened with the green of moss, while at the base, on the side that presumably faces the prevailing wind, a cavern suitable for a family of wood mice. There a small black beetle clambers its way into the safety of darkness, the edges of the cave wall broken and crumbling, each ring of age slightly separated from the next. It reminds me of Fingal's Cave, on the Isle of Staffa in the Hebrides, the broken structure of the wood standing like the basalt columns that march from the Scottish coast across to the Giant's Causeway in Northern Ireland. Then the smell catches me. A deep and damp, earthy must that seems incongruous with the heat of the day.

I walk on, passing another mossy pillow that should start producing chanterelles in another few weeks, and then to the bank

– the leading edge of a wedge of earth that cuts through the forest like the bow of Eggardon Hill. Here, as expected, are dozens of dots of yellow. Little mushroom blobs that look rather pitiful among the dust of a hot July day. I rarely take any from here; the graininess of the soil seems to hinder growth and dirty those chanterelles that do grow big enough to pick. I like to check on them, though, a reassurance that they are still there and that the other spots will also be productive. As I climb the bank and on to the plateau, the beeches give way to birch. Trunks tightly packed and flaking silver, the mulberry-tinted branches forming an almost impenetrable mesh, with an occasional spruce or holly adding coniferous green. I won't head far into this part of the forest. I have explored it, but the ground is so thick with ferns, fallen branches and rotting boughs that I don't like thumping my weight down upon it. This is the naturally occurring woodland habitat that cannot be created without time and it is thick with life; each footstep could crush a microcosm. The trees are busier places too. As I listen, the air is full of seeps and sharp whistles from goldcrests and a variety of tits. They are difficult to spot, so thick is the foliage, and the noises are so soft they are impossible to pinpoint. I spend several moments enjoying the subtlety of sound and the change of air. The breeze brushes the very tops of the birches but this pocket is shielded by the rise of the forest and instead the cool but clammy air from the ground gently kisses my skin. I am wearing shorts and a T-shirt, but my feet are hot inside my boots and the straps from my bag and binoculars feel tight upon my shoulders, squeezing my skin which sweats in response. The humidity is adding to my mild discomfort; I need to move before I slowly melt. Back down into the beeches where my footsteps are less destructive.

I mentioned earlier the notion of trespass and an individual's responsibility to respect ownership of the land. And as strange as it is to consider an ancient landscape three-quarters covered in trees, so too is the notion of 'free land'. To tread wherever you choose. Perhaps, though, in the twenty-first century liability is as great a concern for a landowner as theft or vandalism. If you have a footpath running through your land and an individual errs from it before having an accident as a result of work you were responsible for – a hole half-dug for a fence post or a loose piece of barbed wire in the grass – there might be serious ramifications. Maintenance of a footpath is the responsibility of the Highways Agency or the owner of the land upon which it is sited, and incidents deemed to have been due to a lack of care can be legally compensated. Common sense and personal burden might suggest that an individual should consider themselves responsible for their actions, yet the law, rightly or wrongly, does not always concur. And modern culture with 'no win, no fee' offers has created a rather murky society. Another interesting angle to consider is how much private ownership originally came about. Land was effectively grabbed, those first landowners spotting the opportunity to exploit, creating ownership by threat and aggression. This led to the feudal system and class hierarchy, something that remains in place to some extent today, especially in rural areas. With this in mind, it is understandable that people are so passionate about their right to roam – if the original ownership was likely gained through nefarious means, their claim to rights carrying some moral weight even if time has taken but a few ounces of it.

And my thoughts today of rights and claiming are not without prompt. This new month is *Elembiuos* – the Moon of Claiming. A time when the tribes gathered together for Lugnasad, a harvest feast held to honour the god Lugh. This is a festival marked upon the Coligny calendar, but Xavier Delamarre links the month in which it falls to a deer – *elen* representing deer in Gaulish with links to Greek (*élaphos*) and Welsh (*elain*). This in itself is a little conflicting. If *Elembiuos* is traditionally the time to hunt deer, then it would likely not fall in summer. This time of year is when deer are most difficult to see, with the majority of plants and trees at their peak growth. It could, perhaps, be linked to the cycle of the roe deer, which is one of only two species (along with the red deer) native to the British Isles. Roebucks rut during July and

August, with the doe giving birth the following June. Perhaps, as speculated in *Equos*, the month might be so named as to recognise an animal that was regarded highly in Celtic culture.

Caitlin Matthews, however, links the month to Lugnasad ('Lughnasadh' as she spells it) and a time when 'marriages are contracted and when cases are presented to judges'. A time then when a man might go out and claim a wife. What is confusing here, though, is the timing. While Lugnasad appears in the Coligny calendar within the month of *Elembiuos*, it is also recognised to occur on 1 August. Similarly Imbolc, Beltane and Samhain all fall on the first of the month, yet the Julian calendar is solar-based. The August date might well argue the positioning of *Elembiuos* within the solar year, but were these festivals accurately placed when transferred from lunar to solar?

And now comes my personal crux. I set out today to walk within the Moon of Claiming, that being the assimilation that I felt most comfortable with. Principally because among all of those lists that I have found, Claiming appeared alongside other moons that seem most unambiguous in terms of etymology. Having settled upon the *Samonios*/Samhain link, I rather let the rest of the pieces fall accordingly without an enormous amount of consideration on my part. I agreed with that fundamental point and so assumed that my thoughts would align accordingly. The link that I imagined I would find today was a claim of right. I would be picking a few mushrooms upon land that isn't mine. This sounds naughtier than it is because I have spoken to the manager of these woods and got 'unofficial' permission to forage (one of those 'at your own risk' conversations that links to that issue of liability), but it might, I thought, have tied in rather neatly.

As I have walked, though, so my thoughts have shifted. Being among the trees, having a purpose but also without pressure of time, I have touched upon my connection with the woodland. Those things that matter to me, or more specifically those things that inspire me, are often from the natural world. My year is dictated by events that occur at certain times, encouraging an association that is unshakably profound. If I was tasked with naming the months of the year then they would probably reflect the natural course of things. And if my life was intertwined with a certain species then naming a month after it would be less a tribute than a simple recognition. Deer were vitally important to the Celts, especially in the British Isles at a time when rabbits and brown hares were not here as a source of food. While I walk below these beeches then, I am erring towards a different association to the one in mind as I locked the car. It will be interesting to see whether that direction holds, when I am back at home, reading another potential source and detached from the present.

I have stumbled upon a slightly unexpected bounty. I hadn't come to this part of the forest before today because it doesn't usually produce chanterelles until mid-August, and then many get dusted and dirtied like the ones I saw earlier. But there has been a change since last autumn – nothing major but enough to prompt a rather different scene.

I had left the birch trees to check a single beech that must be the most mycologically productive tree in the forest – in terms of chanterelles at least. As always, the ground surrounding the base of the trunk is absolutely studded with yellow. I have never tried to count them, but there must be several hundred chanterelles that push skyward around that tree each year. My next pause is in a triangle of trees, edged on one side by a bank that rolls over itself like a wave on a swollen sea. There, early in the season, comes an enormous flush of ceps. They are a mushroom more highly regarded than chanterelles, known in Italy as porcini and in Britain as penny buns. The first big dump of rain that falls after the beginning of the final week of July will cause the explosion, and then there is a race to reach them before the slugs ascend. The slugs time their appearance to coincide with the ceps and, given how good a fresh cep tastes, it's hard to blame them.

I walk on, crossing another, much shallower, ditch before climbing up a slope that looks as though it will roll on forever. The point I am at now is a bank around 12 feet high, and all I can see above is the shelter of beeches and the brown of dead leaves that scatter the forest floor. There are no birch or spruce here and very little understorey. Instead, the beeches have spread their branches and filled every inch of space, like a great march strolling shoulder to shoulder. This is possibly an older part of the forest, perhaps an area where the trees avoided felling in the 1940s. They certainly look big enough, the trunks wider and the bark a deep, weathered grey, with knots and notches that look like scars on the hide of an elephant. The atmosphere is nothing close to that which I felt in the Forest of Dean, though. There I experienced a sense of insignificance not dissimilar to that of a Scottish mountainside. A presence from something living that is a sum of all of its parts. Humankind isn't welcome because humans remove that cohesion – the equilibrium of ecology that we have long since ceased to be a part of. There is much to be gained from walking in these woods, however, not least because I suddenly have enough chanterelles in my basket to share a few around. At the bottom of the bank is a damper area and a scratch of pine and holly, a barrier against the might of the beeches above. A couple of the taller pines obviously succumbed to a winter storm and as they fell they took several branches of beech with them. As I approached, my heart briefly sank. In a further moment, I felt a sense of disorientation – perhaps I wasn't where I thought I should be? I skirted around the debris and then worked down the slope a little further along where the gradient is less sharp. Those branches had all crashed smack on top of one of the largest chanterelle patches in the forest. Admittedly they are not always high on quality, but nevertheless this would be a distressing sight for any would-be forager. The slope itself produces fewer bodies, but as they spill down so they find their feet in a thick carpet of moss which keeps them damp and encourages 'forced' growth. To cling on to their toes and still reach far enough through the moss to reach the air and spread their spores sees them form fat and long. I pick several, climbing up the bank a few feet to reach one almost beyond my grasp. And then I notice a great dollop of yellow beneath the fall above.

The broken limbs had not, of course, fallen flush upon the ground, and the mycelia, having suddenly found a roof above their heads, sounded the alarm. Wherever I could peer or work a hand I found great big dreamy chanterelles. I could be fussy, picking only the finest specimens, and now I have a haul that would impress a Michelin-starred chef and I've barely dented their number. Millions of spores will still be cast but this season is not looking as bleak as I feared. The sun can beat hard this next fortnight and the temperature soar, but while the chanterelles shrivel elsewhere on this mycological pathway, this secret cache should endure until autumn.

8

THE MOON OF DISPUTE

I've watched you now a full half-hour; Self-poised upon that yellow flower
And, little Butterfly! Indeed I know not if you sleep or feed.

(William Wordsworth, 'To a Butterfly')

I DO WALK in the rain. This might seem unlikely given the weather conditions that I continually relate, but I will, honestly, step out of the door when the sun isn't shining. In fact, if I'm fishing, I welcome the dampness, especially those dank days of winter when the mist and rain roll into one singular mass of moisture. Those sorts of conditions seem a lifetime away today, though, despite the fact that the days are noticeably shortening. The sunshine is unbroken, although a bank of cloud is working slowly towards me from the north-west. The breeze is also northerly, but with an easterly bite which is converging with the cloud and stalling its direction. Along its edge it has been buffeted upwards and back on to itself, like water in slow motion colliding with an invisible wall. I open the gate, step through, and walk across the meadow to the entrance of Eggardon hill fort.

A month has passed since I ventured up here on that sultry evening to be blown away by the sunset. The temperature is considerably lower now than it was that evening, but I remain in my shorts and T-shirt, with wellies on, as ever, to combat against ticks rather than negate the mud. And the ground will be soft in places, this fine day not reflecting the rest of August. Since the scorch of July came to a thundery halt, the weather has been typically unsettled, a flip of the jet stream and a procession of weather systems. Many of the fronts have been weak, like the languid fold of cloud today, but there have been a few drops of rain and plenty of puff.

The people keep coming regardless. The caravan parks spilling into the spare fields that are normally only used on a sunny bank holiday weekend, while an increasing number of farmers have made room for makeshift campsites. There are a good number of cars parked in the layby today, but I need to get out and look for something that will make me catch my breath. Something that will snap me out of melancholy.

The rise of populism in Britain, and the US, has led to one of the less savoury political passages, certainly of my lifetime. Ever since we reduced one of the most complicated of issues into a two-box decision in 2016, it seems that everything has become binary. The majority of people didn't care less about EU membership before the referendum, but battle lines were drawn and two sides began their tribal tubthumping. And the hangover is worse than the event itself, an incumbent government winning a landslide on a campaign built on emotive division. People weren't voting for policy but cheering their favourite wrestler to the ring, riled in part by the 'told you so' barbs coming from the opposite side. Nobody debates any more, they just throw insults or parrot meaningless slogans and epithets. And more telling is that no one listens or tries to understand an opposing view. They just dig a deeper hole behind that first mark and keep chucking grenades over the top.

I could easily avoid it all. Social media is the root of so much discord, a place to create a personal echo chamber or level offence without recourse. The initial purpose has been swamped by marketing, misinformation and personal gain. People make a living from being contrary and controversial, the audience that always bite back justifying their objector's worth. Yet it can also be a positive place, filled with humour and wonder. Somewhere to track down a source rather than a filtered opinion, or consider the view that you might not share. And although I am somewhat ashamed to admit it, watching people bicker or stumble into comeuppance is compelling.

Switching off is not quite so easy as it should be. Those small screens are horribly addictive and I reach for my phone just as I once reached for a cigarette. But there comes a point when the baggage begins to weigh heavy. Eighteen months of restrictions and news saturation are taking their toll. People are, understandably, fed up with the pandemic and the constant noise around it. And it doesn't help having a populist government who test the water and drip-feed the media before deciding upon a plan of action. You can't please everybody, but you do have to be honest.

When I was a child learning about history and those mistakes or events that seemed so preposterous and glaringly avoidable, I had no idea that we as a species hadn't learned from them. Just as every generation must look into the past with a frown and mistakenly believe that they, at least, were living in a time of learned retrospect. These last few weeks have found me in a spiral of my own making, induced in part by self-doubt as I write this book. I am, and always have been, a master of procrastination, and to snap myself straight I need to see something extraordinary.

It would appear that this time of year has a long association with discontentment. *Aedrinios*, the Celtic month in which we are now in, is widely interpreted as the time for dispute, but this, as ever, is rather ambiguous. Delamarre linked the name with *aidu*, the Gaulish word for 'fire', and this could certainly link metaphorically with the common theme, although the relevance is dependent upon the corresponding solar timeline. Were *Aedrinios* to fall in winter then the connection to fire, in terms of its importance for survival, would have substance, but in a time when winter was deeper and more prolonged, it would have related to more than a single month.

What then might a dispute at this time of year constitute? One immediately thinks of war, but that level of conflict is not something dictated by the passage of the moon. If people were open to the concept of waiting until a specific month to pick up arms, then perhaps those arguments were not worth beginning in the first place. What seems more likely is that now was an opportunity for a moot meet: an assembly formed to settle civil disagreements, perhaps the debts held over from the previous year that can now be gauged against the likely yield of the current harvest. Borders and territories might be bartered, dowries and future unions agreed. The assumption of something more barbaric likely stems from Roman and Greek portrayal, but knowing how civil the Celts actually were (again, I should clarify that human sacrifice would not fall within my personal scope of 'civility'), how they traded and engaged, 'dispute' is likely a term with positive connotation. Promises can be made before all hands return to the fields and the harvest gathered. Such a time would require total cooperation so anything that might jeopardise that harmony should be resolved beforehand. We can never know fully the laws by which the people who lived here, upon Eggardon, might have lived, but it is reasonable to assume that this was a society of structure. Certainly more harmonious than the Romans would lead us to believe.

My walk to the first gate takes me past a couple of tumuli. These circular mounds are found across this country and much of the world and are usually burial locations. The field in which those two are situated is managed for pasture, but the mounds themselves are never touched. Cattle and sheep will still ruminate across them, but their original purpose has always been respected. As a result, the flowers and grasses that grow upon them do so with a little more vigour than elsewhere. The greens, purples and yellows are deeper and denser, the tumuli even more conspicuous as a result. When they were first created, the soil disturbance would have led to a rapid germination of dormant seeds – rather like the grass-snake bank that bloomed so brilliantly back at home. For the Celts, this would likely have represented a portent of passing and perhaps the reassurance that the path to the next world was peaceful and successful. Subsequently, the degeneration of the bodies would have provided nutrients for the soil, furthering the lush growth upon the top. It is rather nice that the process has endured to some extent. Here at least. Elsewhere, tumuli would have been ploughed in or simply sunk to nothing, only revealing themselves in times of drought or through geophysical research.

It is no surprise that the dead were buried a slight distance away from the main body of the fort. On a practical level there would be the risk of disease and, depending how deep the tombs were dug, an unpleasant smell. Symbolically, the souls might then be free to leave the confines of the tribe and find an easy journey onwards. My Celtic musings continually deepen as I spend more time walking here. Much of my supposition is built on common sense or learned knowledge rather than personal insight, but there is no coincidence to the fact that my understanding becomes more profound when physically present. It isn't always easy to tread in the footprints of others, but it is harder still from the comfort of familiar surroundings. The lack of care and actual hate for those people trying to cross the Channel in rubber boats is heightened by detachment. Were those detractors to actually witness the desperation, to see people forced to leave their homes and families before placing a final hope in a tiny, overcrowded boat on a volatile and hazard-filled sea, perhaps they might show a little more care.

The footpath takes me diagonally across another field of pasture, the grasses thick and deep, lush green. This is improved grassland, managed and planted with ryegrasses that are fast growing and ideal for fodder. It was cut a few months ago, so all this growth has come since, and will likely be cut again before the summer is over. The lack of variety and absence of flowers means there are few pollinators to be seen, but crane flies are abundant and every few paces I seem to disturb a small clutch of linnets, their lively chatter sounding like the rapid, repeated squeak of a dog toy.

To my right is the eastern rampart of the fort, running adjacent to the edge of this field. The seed heads and stalks of the grass upon it are fading into yellow, far from the dusty scorch that might come in late summer but tired, nevertheless. But there is colour among it, splodges of yellow and mauve that I can only see if I don't look directly at them. Something else catches my eye, though: a small bird that has appeared on the top of the mound. I know what it is before focussing the binoculars upon it, so distinct is its pose. Upright and confident, head held high with the slope of the back running neatly to the end of the tail. It is the bird I come looking for up here in spring, when its appearance brings a different kind of emotion.

The wheatear is a bird of the uplands. Wintering in Africa before heading north to breed. They are small birds, bigger than a robin but not as large as a skylark. The males have dark wings and a blue-grey back and distinctive black mask. The females are not quite so distinctive other than by the feature which both sexes share. The tail of both is edged in black with an inverted 'T' that slots neatly into the brilliant white of the upper tail and rump. It is that flash that is so conspicuous when the birds take flight, easily distinguishing them from other small passerines, such as pipits and chats, that they might share habitat with. It is also believed to be a source of their name, coming from the Old English for 'white-arse' which does link to the French common name *cul-blanc*, the slightly less rustic 'white-rump'.

I love to see them in the spring. They are often the first of the long-distance migrants to arrive, some birds dropping in before heading further up into the Arctic Circle while the moors of the south-west and hills and mountains of Wales, the Peaks and anywhere further north in Britain might support breeding birds. In Dorset, a pair might occasionally breed. They have done so on Portland in recent years and recently fledged juveniles were spotted local to here a few summers back. They do seem to loiter on Eggardon with the intention in mind, and it is classic wheatear habitat. The drawback, and almost certainly the annual deterrent, being the volume of human visitors. A sunny Easter weekend here will find a steady stream of walkers and kite-fliers working on and off the hill, just at a time when any wheatears might fancy setting up a territory. The birds seem less wary as they head south, a trait perhaps distorted by the presence of juveniles who might not identify humans as a threat. They are also agile on their feet, often running to catch the insects upon which they feed. Such mobility might give the wheatear greater confidence to evade danger than were it relying wholly upon flight. Seeing them at this time of year is bittersweet, though. A reminder of the change in season, rather like watching swallows line the powerlines until one day they have gone.

I press on, through the next gate and over the stile. A couple are picnicking on one of the inner ramparts and I smile an acknowledgement before cutting a direct path up and down and up again, using a sheep track to find good footing before stepping up on to the highest lip that runs around the central plateau of the fort. There are more people milling, some walking the ridge further along from where I stand and a young family with three children who shriek and skip excitedly. I have a cursory scan with my binoculars but do not expect to see much. On a quieter day there might be half a dozen or more wheatears hopping and perching across the plateau, with a chance of yellow wagtails if the Highland cattle are present. It hasn't been grazed for some time, though, the sheep and cows on lower slopes, and the grass too tall for a wheatear to feel comfortable in. There are a few flittish goldfinches working the thistle heads, but little else, and the breeze across the top is stiff, as ever. Too strong for a butterfly to fly against with any purpose, although the direction should mean that I will shortly find plenty, in an area I don't often see them.

As the trees chattered in my thoughts on my search for chanterelle mushrooms, so I imagined the landscape as it might have been thousands of years ago. When the fort here was created, the surrounding area would have been dense forest. The wind might have checked the growth on the exposed rise, and ruminants would have kept some areas clear of growth, first as wild animals and later domesticated and tended. But the trees offered plentiful resource and would have been well used, for fuel, housebuilding, fortifications, partitions, tools and weapons. The cleared areas were then grazed and farmsteads created, but deforestation was a relatively slow process simply due to a smaller population requiring less.

I pause briefly, looking down to the sweep of land that curves away to the south-west. The slope is steep and marked with terracettes, a series of evenly spaced ridges that look rather like a grand flight of steps. They are widely believed to be formed by the habits of grazing animals, particularly sheep, who graze horizontally and, over time, leave their own impression. There is some doubt to this theory, but there is no question that it suits the sheep that are present today. As I watch, they work slowly sideways, at different heights and distances to one another, but all sticking to the terracette upon which they find themselves.

Were they maintaining the exact same pace, and were there not a few lying around enjoying the sunshine, then it might look like the sheep are powering a giant treadmill. An endless, methodical plod as the world turns beneath their feet.

If the terracettes are indeed formed by the feet of grazing animals, then that slope would have been much smoother and harder to climb at a time when the fort might have required defending. And if it was originally wooded, then it would surely have been cleared for that very benefit. There does not appear to be any evidence that Eggardon hill fort was ever attacked, but archaeological finds at other hill fort locations certainly point that way. Stashes of sling stones have been uncovered at Danebury in Hampshire and mutilated body remains found at Bredon Hill in Gloucestershire. Yet, as more of the wider landscape was farmed and utilised, the need for a singular fort as a site of settlement would have been questionable. It might have been hierarchically inhabited or prepared as a place of retreat should a threat be forthcoming. Archaeological excavations made in the 1960s on the earthworks here on Eggardon point to the mid-Bronze Age, so more time would have passed between the fort's construction and the arrival of the Romans than has passed since. A sobering thought in many ways, not least when I consider the impact that humankind has had on this landscape in the more recent past. It also shows the folly of attempting to pigeonhole the functions of this place. With so many years of existence, of human advancement and societal evolution, the fort as it was 2,000 years ago would have been almost unrecognisable to the people who first dug the barrows and ramparts.

That curious concept of time and its passing: how it drives us to bookend and surmise. And should we question something, then it is too easy to believe that further ahead will come more definitive answers – technology and learning will prove the things we doubt. Yet as time passes so the evidence from which such efforts might be based is disappearing, and our connections to the past, those gossamer-thin threads that modern society seems intent on breaking, will one day vanish altogether. Of course, what could, and probably will, happen is a discovery or two that offer complete contradiction to what we currently understand. Another Gaulish calendar might be uncovered, complete this time, or one found elsewhere in the lands of the ancient Celts. It might hold keys that the Coligny calendar does not. Just as people today are rethinking the interpretations already made, so their theories, no matter how sound they seem on the evidence available, are turned on their heads.

The reality is that we will never know the answers, and I take some comfort in that. I can feel connections and beliefs that might have some basis within learned knowledge but are also influenced by my own thoughts and emotion. And that is a nice way to view the world.

There isn't the intensity of colour that I found here on that sultry July evening, and that incredible smell is all but absent, but still the grassland looks resplendent. Much of the clover has gone to brown, but the bird's-foot trefoil still glows with the sun and the blue-bowed heads of harebells add a sense of demurity to the scene around my feet. And I am careful with each footstep, wary of squashing both the flowers and the thick buzz of pollinators that are feeding upon them. There is a crumble of a path down the slope and I try to use it when able, but it is a route created by the cattle who obviously trod it when the ground was wet, as their deep hoof-marks have dried and left the surface pockmarked like a toasted crumpet. And as I step out of that fresh northerly breeze I wonder if I might melt like butter on that crumpet. I have walked into a suntrap and the extra heat is marked. In a moment I am whisked away to my favourite place in Scotland – in the whole world in fact – and an extraordinary invasion of painted ladies.

On the final Friday of May 2009 I was last out of the office – tying up the loose ends before Sue and I headed up to Mull for a fortnight. As I locked up and wandered over to my car I was showered by a fall of butterflies. I stopped still in the early evening sunshine and looked upwards, half-expecting to see a giant bucket – tilted and overflowing. Every few seconds a clutch of around a dozen painted ladies would burst from apparently nowhere, before slowing and spreading as they reached head height and dispersing around me. By the time I got home they seemed to be everywhere, feeding hard on every buddleia in every garden. It was a pattern repeated across the south, as millions of butterflies made landfall having flown from the desert fringes of North Africa. The following morning, as we headed north, the painted ladies seemed to be matching our progress, appearing each moment we paused for toilet or lunch breaks. Eventually, having reached the Isle of Mull and met with my parents at our family's long-time favourite holiday residence, I walked around the bay and sat on a rocky outcrop. Breathing in the sea and breathing out the stresses of the weekly grind. As I did, a painted lady fluttered across the surface of the sea and settled briefly on a rock face beside me.

That was a particularly special fortnight. One of countless that the Parr family (with a variety of friends and relatives) made since the mid-1980s. Most of those trips were accompanied by typical Western Isle weather. Unsettled from one hour to the next. But for those two weeks the sun shone throughout, and Carsaig, the hamlet where we stayed, enjoyed its own microclimate. Much of the fine weather was accompanied by northern or easterly winds, but with a south-westerly vista and 1,000-foot cliffs as shelter, Carsaig roasted. The temperature on the northern edge of the Ross, just 3 miles away, might be 10 degrees Celsius chillier than in the suntrap of Carsaig, a similar effect to that which I am experiencing today.

Here, on the southern slope of Eggardon the ground seems to be gently cooking. This spot, beneath an almost sheer drop of 50 or so feet, is feeling not a whisper of that northerly wind, and, without a breeze to bother them, there is a billow of butterflies. I gave little thought to the butterfly potential when I first walked here; I would see the occasional flutter, but I would normally be strolling the higher points where I had more scope. The habitat is perfect for many species though: well-drained, chalk grassland that is grazed but never to excess. The only drawback is the wind, but that doesn't mean the butterflies aren't here – they will just be sheltering. By walking the ditches, particularly on the leeward side dependent on the specific wind direction, I began to find serious numbers. What is unusual, though, is to find such numbers on the open southern face – feeling the full warmth of the sun and with food plants that don't normally get visited. And, rather appropriately, just as I have reached level ground and taken stock of my immediate surroundings, I notice a painted lady just a few feet from me.

It is feeding on the lilac-mauve of a thistle flower, and for a moment I think it is a red admiral. I am more used to seeing painted ladies perch with open wings, but this one sits in profile to me and, being at eye level (the thistle sits on a slightly raised bank), the low, late-summer sun is lighting it up with an unfamiliar glow. There is an extra depth to the orange of the upper wing, so rich it nudges me towards the red of an admiral, but judging by the clean, fresh lines of the rest of the insect, this is a recently emerged butterfly in its finest livery. The lower underwing is captivating. I cannot recall having an opportunity to study a painted lady in such a manner and I had certainly never appreciated the olive-green mosaic before now. It seems to reflect the deep colour of the thistle leaves that stand like spiky fronds of seaweed in the still of a rock pool.

Given the measured drop in butterfly numbers over the last few decades and the shifting baseline that likely preceded record, it is more than reasonable to assume that these slopes were aflutter even when the fort itself was in full use. But butterflies and moths do not seem to play a role within Celtic symbolism, which, although not surprising given the lack of direct engagement people would have had, is perhaps curious considering the marking of time. Many species appear for relatively short periods to coincide with their favoured food plants, and, as a result, would tie in fairly neatly with the Celtic calendar. Perhaps, though, therein lies the value of awareness at a time when survival would supersede study. The animals that were clearly revered existed alongside people for food or work in the case of species such as horses and deer, or specific totemic symbolism. It would make more sense to honour an eagle for its enormity and awe rather than a humble and ubiquitous butterfly.

As previously mentioned, humankind's deeper interest in ecology is a very recent thing. Gilbert White was a pioneer as late as the eighteenth century and some species of butterfly were not identified until long after his passing, such as the Lulworth skipper (in 1832) and Essex skipper (in 1889). Unlike the vagaries and supposition surrounding anthropological history, we are always learning and discovering new things about the world around us. Our knowledge increases with the passing of time and understanding the intricacies of life cycles can aid conservation. The large blue butterfly was declared extinct in Britain in 1979 but has since been reintroduced – that success due in large part to a greater knowledge of the species' symbiotic relationship with the red ant species *Myrmica sabuleti*. The female large blue lays her eggs on wild thyme, with the larvae subsequently feeding on the flower heads and developing seeds. When they reach around 4 millimetres in length, the larvae then fall to the ground and produce a honey-like excretion that attracts the attention of the ants who carry the larvae back to their nest. There, the larvae are placed in brooding chambers where they feed upon the ant grubs until they are able to pupate, eventually emerging as adults in early summer.

Identifying such a process requires an immense amount of fieldwork and an open mind. It is not always easy to observe what is actually happening without seeing what we think should be happening. Why else would a red ant carry off a butterfly larva other than to eat it? The idea that this was a process to benefit the butterfly would take a depth of perception, something that we as a species are not always prone to develop. In detaching ourselves from the natural world so we place ourselves above it. An unpleasant arrogance wherein we view other living entities as subordinate playthings. Unless something has a 'value' to humankind it is expendable. We are told to think twice before squashing a wasp because they are, in fact, useful to us – they eat aphids and other invertebrates that might damage crops or gardens. But we shouldn't need a reason at all. Our duty to the world around us is not to decide what has a right to life and what doesn't, but to maintain an environment that gives every species the opportunity to function and exist as it needs to. After all, who might have suspected that a lack of red ants could have such an impact upon the large blue butterfly?

Much as we try to disassociate ourselves from the rest of the natural world, we cannot escape the similarities to it. In fact, humans are not much different to the red ants in terms of basic societal structure. We each have our roles and struggle to fill the positions to which we are not suited. Some of us are leaders, but far more of us need to be led. My own self tends towards pleasing everyone. I could never lead successfully but neither will I blindly follow. Instead, I am a mechanic, intent upon creating harmony and avoiding those tough decisions that a stronger mind can make. I am happiest when people get along and find confrontation difficult, especially for myself. Arguments or cross words, even via email or social media, leave me scarred and vulnerable. I have always gone out of my way to avoid physical conflict, although twice I have found myself in the wrong place at the wrong time. Where unashamed pleading was met with random mob brutality.

The first occasion was on a south London street, walking to a college bar with my then girlfriend and getting attacked from behind by three men. They kicked and punched and ripped at my clothes, the solid stitching of my jumper saving me almost comically as it stretched and then gave suddenly, forcing the guy tearing at it to stumble like a domino into the other two. I took the chance to run and escaped with only bruises and mild concussion, but the mental scars cut deep. For many years I was afraid of my own shadow, avoiding darkness and wary of everyone. On a train ride back up to London I twice got off to catch a different service, just because I was in such an uncomfortable state. The problem was, I knew it was going to happen again. It wasn't paranoia but a deep-rooted acceptance of something inevitable. And when it did happen again, in Winchester this time and in equally unprovoked circumstances, I subsequently felt a sense of relief. The physical pain was largely superficial once more, despite having a glass smashed in my face and a week of concussion as a dozen excitable teenagers used me as a punchbag, but within a few weeks there came a sense of clarity. I would take nothing for granted and would remain painfully aware of myself, but the feeling of something foreordained had gone.

I leave the painted lady and turn to the steepened face that is bathed in the late-summer sunlight. Almost immediately I catch the movement of blue. Then another. And another. These are not large blues – as yet there are no colonies in Dorset – and neither are they the species I had hoped to see. But they are striking nevertheless. The common blue butterfly is a familiar sight and one of our most widespread species, occurring right across the British Isles with the exception of Shetland. They are fond of bird's-foot trefoil, so it is unsurprising that they are present in good number on Eggardon. They are not the most frequent species here, though. The small heath, similar in size to the common blue but orange-brown in colour, seems to be the most numerous species today, while a few weeks ago the meadow brown, our most common species, would have dominated the grassland. There are still plenty of meadow browns across the hill but they are looking washed out and weathered, like a T-shirt faded by the sun. Their main flight here is over and it will not be long before there are none. Other species, such as the small heath and common blue, will have more than one brood each year, perhaps giving the appearance that the adults are far longer lived than they actually are. It is a similar strategy to some bird species – those particularly susceptible to poor weather. Stonechats, for example, might collapse in number after an inclement year so will attempt to raise multiple broods in subsequent summers. The small copper, one of my favourite butterflies, will do similar, with individuals sometimes visible from April to November across four different broods. They are a species I expected to see today, and it doesn't take long to find one.

I have stepped away from the main face that seemed covered in common blues and along into one of the wide ditches that run between the ramparts. The sides are steep and smooth, but wide enough for plenty of sunshine to reach into, and it seems that every flower of every species is feeding a pollinator of some description. Hoverflies, bees, bumblebees and a throng of burnet moths. And on a dandelion beside my feet sits a freshly minted small copper. It is small but showy, wings spread and the copper of the upper wings glinting and glistening in the sunlight as though buffed and polished. The black spots on the wing are ink-blotted and jet, while even the duller brown of the lower wings seems to glow. The males and females are similar in appearance but judging by the small size I presume this one to be male. He seems unfazed by my looming bulk, although I did inch towards him with care and made sure I didn't cast a shadow across him. I pause and watch him for several minutes, the air in this dip cooler than against the face, the sweat that had begun to cling to my forehead and neck gently receding in the comfort. He is a glorious insect and as I look along the gully I see two more. But the small copper is not the butterfly I had hoped to see. The butterfly that might shock me out of the doldrums.

I had taken a few photographs of the smarter-looking common blues that were flying around the sun-baked face, not just to admire later on, but to double-check exactly what they were. In fairness to myself, a freshly emerged common blue can look spectacular, but after a few further paces I am confronted by a dazzle of colour that cannot be mistaken for anything else. This is a male Adonis blue, considerably less widespread than the common blue, but when fresh, as this one is, and bathed in sunshine, it has a vibrancy quite unlike any other British butterfly. It is feeding upon what looks like field scabious (I don't want to peer too closely in case I spook it) and as it moves around the flower, tongue probing, the wings flash in the sunshine, the tone of colour changing dramatically with every glint. It is reminiscent of the ChromaFlair paint that is used on some expensive sports cars, an iridescence that changes depending on the angle of the viewer and light source. Like the star Sirius on a crisp, cold night. Such a quality can never look as good as it does on a butterfly wing, though, and the difference to the common blue, as lovely as they are, is inescapable.

I have to smile at my own uncertainty, and recall the exact same reaction around twelve months ago. And I am reminded of a day nearly a decade ago when I visited the Isle of Mull with my friend Dan Kieran and we ventured out with another friend and local guide Bryan Rains to try to spot an eagle in the snow. Dan had thought he had seen eagles in Scotland before, but the views had been distant and he wasn't wholly convinced. Despite the mid-December weather, Bryan soon provided us with gold. The eagles were sitting tight on far-away crags so, although unmistakable through a telescope, their size was tricky to gauge. Then a white-tailed eagle loomed over a low ridge behind us, the great 8-foot wingspan looking even more imposing against the white of the snow. Dan's eyes bulged and his jaw dropped. He turned and raised an eyebrow. 'Okay, I have *never* seen an eagle before ...' he said.

It is easy to lose perspective when in the enormity of the Scottish mountains, and the buzzard, similar in shape and colour to the golden eagle, is often referred to as a 'tourist eagle', due to the fact that so many eager eyes make the mistake in identification. As is often noted, though, if you are in any doubt whatsoever, then you are looking at a buzzard not an eagle. That same sentiment rings true at this moment. There really is no doubt that I am now looking at an Adonis blue.

And the effect is as I hoped. For several minutes all other thoughts vanish from my mind. And even though I thought I knew what to expect, I still wasn't prepared for the splendour. My imagination could not come close to recreating the spectacle itself. And now I am buoyed with a quick shot in the arm and open-mouthed wonder. I will smile until bedtime.

9

THE SINGING MOON

Voice of summer, keen and shrill,
Chirping round my winter fire,
Of thy song I never tire.

(William Cox Bennett, 'To a Cricket')

THERE IS NO DENYING that I have dug a bit of a hole for myself this year. It is naivety on my part, an assumption that others would have long before determined the details for me to whimsically dip into. Perhaps I should have made a tactical withdrawal early on and diverted my interest to the Romans. There is, after all, a Roman road running along the ridgeway opposite me and no obvious signs of Celtic activity visible from the lounge windows. But much as they have influenced this land, they have not shaped its soul. They were here for less than 400 years, in which time much of what is currently England and Wales came under their rule, but Caledonia and Hibernia (Scotland and Ireland) remained out of reach and unconquerable, Hadrian famously building a wall to the north while the west was a sea too far. And after they left, the Celts remained, settling in the Roman-built towns rather than returning to the hill forts, although that may have been a natural progression anyway, influenced by their neighbours on the mainland of Europe. There remained little impulse to keep records of their lives, though, and as Britain slipped into the Dark Ages definitive knowledge disappeared too. A reason why Celtic mysticism endures. Just as it cannot be truly determined, neither has it ever truly gone.

Mind you, I would have found my hole pretty boring if I stopped digging the moment I unearthed a mosaic of Roman tiles and a few loose sestertii. And the fact that I will find no comprehensive reference has made the journey that much more absorbing. The consideration of something can be just as fulfilling as the knowing. Liberation comes when we stop seeking the final answer but keep asking the questions – particularly of ourselves. And arcing back to the T.S. Eliot poetry at school, the vital aspect is feeling justified within my own mind. Something which is a little bit tricky with the moon that will rise tonight.

At least I can begin with something that all minds seem to agree upon. This month, *Cantlos*, was the final full month of the Celtic year. Whether it fell around this point of our solar year will ever remain in debate but, as I have concurred with the notion that *Samonios* links with the festival of Samhain and 'summer's end', *Cantlos* must then precede it. The etymology of *Cantlos* doesn't help with its station, though. It is hard to argue against the translations made by Delamarre and others, linking it with the Old Irish *cétal* meaning 'song or recitation', and the Welsh *cathl* 'song or hymn'. The Latin word *cano* means 'I sing', as does the Proto-Indo-European root *kan*, from which chant and incantation are formed. All of this points rather decisively towards *Cantlos* being the 'time for singing', which is suitably ambiguous to fit almost anywhere within the Julian calendar.

There is a small niggle in my mind that a 'time for singing' could relate to the peak of territorial birdsong, which would fall in spring and allow *Samonios* to straddle the summer solstice, therefore supporting the argument that the Celtic year did in fact begin with the longest day of the year. This seems unlikely, though; I made the point previously that the Celts were unlikely to honour by naming something that wasn't intrinsically linked to their culture, while the translations definitely point towards ritualistic singing or chanting, not the fluidity of a blackbird or nightingale – no matter how wonderful they might sound. So I can argue myself out of that quandary, although there remains another somewhat contentious issue. As I write, the moon is three days from full and we are four days away from the autumn equinox. Everyone knows that the full moon closest to the autumn equinox is the 'Harvest Moon'. It is surely the most well-known and oft-discussed full moon of the year, and tying in so neatly with the equinoctial cycle means its identity is undeniable. Except I'm erring towards ritualistic song rather than the sweep of a scythe, and, as I already mentioned, there doesn't appear to be a lot of basis to back that up. There is, however, a case for the rising of a Harvest Moon within *Samonios*, as I will explore, although that in itself will contradict the popular moniker for what is likely the second-most familiar moon of the year. Perhaps I should span that orbit when I get to it. For the time being I have a rather pleasant and unexpected distraction.

I wasn't going to write at all today; instead I had planned to walk at the very point of the equinox beneath a rising moon still almost full. But a stroke of serendipity found me strolling on Eggardon today with an old friend and several new, and sharing the hill with people who had never set foot there before, but had minds and eyes so open to its wonder, was joyous. Afterwards, I felt lifted and a little bit inspired, and compelled to write a little. And as I have done, so has formed a gathering above the eastern ridge opposite the cottage. I'm going to have to leave the words be for a time and wander up and walk among the mass.

The job I left before we moved to Dorset – where I witnessed the cascades of painted lady butterflies – was in a location not typical of an import and distribution business. An office in a converted barn in a Hampshire farmyard, with gable doors opening on to open fields and lunchtime walks on private tracks. Stepping outside offered a genuine form of escape. Much-needed relief from the stress through the glimpse of grazing deer and wheel of buzzards. There was even the occasional goshawk sortie and a purple emperor butterfly resting on the bonnet of the marketing manager's BMW. There were plenty of unconverted outbuildings where swallows could build nests in the rafters, and, on the main farmhouse itself, an absolute mass of house martins. There were at least forty occupied mud cups around the roof, with many more on the old workers' cottages nearby, and by the time the second broods were fledging in late summer the sky would be thick and the nasal chatter almost deafening.

Then, just a couple of years before I left that job and the area, the martins didn't return. There was no slow dwindle or noticeable decline, just an absolute disappearance. I went up to the farmhouse to double-check, walking the outside walls with the woman who lived there. There were no fresh droppings or signs of nest repair, just an eerie emptiness. House martins typically migrate in large groups: a good strategy to combat predation by safety in numbers but a risk to a single colony in extreme weather. And that is likely what had happened. A climatic catastrophe as they headed south to Africa or north on a fateful return.

A little over a year ago we witnessed the potential impact of the weather here. Sitting in the lounge on an autumn afternoon, the air outside suddenly filled with martins. I thought, for a moment, that they might be feeding: perhaps a late flight of ants was offering up an easy meal. But then came a couple of hefty thumps of wind and a sudden drop in temperature. Hailstones rattled the roof and condensation glazed the windows. I opened the patio doors and to the right, along the south-facing wall, house martins were clinging desperately to any available space. Some were grounded, one flopping beside my feet like a waterlogged paper dart. I scooped it up and it was icy cold and barely moving, dramatically downed by the squall. As I held it so I felt its body temperature respond to my own and after ten minutes or so it appeared alert and able. It seemed a little unfair to return it outside but crueller still to hold on to it. As the weather settled and the martins began to take flight, the only hope was that the one in my grasp might fly with them.

Afterwards, I went online and discovered it was a widespread pattern. Short films and photographs from Charmouth and West Bay showed house martins clinging to leeward walls just as they had done here. Many thousands of birds had been downed by the weather, and any that may have been over the sea would not have found sanctuary to cling to. The vulnerability of migrating birds was all too evident and even those that flew off, like the one I warmed in my hands, would have struggled to cross the Channel without flopping.

Today, though, is an altogether different scene. A lowering sun and the lightest breeze. There is a cool edge to the air, but nothing you wouldn't expect to feel in early autumn, and as I lean on the stoutest fence post I might find atop the eastern ridge, my head is in the clouds. I cannot be sure of the number. I am fairly confident with a rapid estimate of moving flocks, but these martins whizz, skim and whirl without pattern. There are many, many hundred, but it doesn't matter too much. Being among them is enough.

I turn to face the small field that the gateway opens into, the ground roughened with flints and loose patches of grass. It was ploughed this year, but neither harrowed nor seeded, and the martins seem to be finding food across it. I turn and crouch, finding a line that they seem to be working. An easy drop for a small bird from 50 feet before a sharp, zigzagged hawk just a foot or two above the ground. They are coming almost straight at me, some birds (that finish the pass with a zag rather than zig) flashing within touching distance. But I stay as still as I am able and they pass me as though I were just another, somewhat more robust, fence post. Judging by the dusky plumage and shallow fork in the tail, the majority of these birds are juveniles. There are a scatter of adults among them, looking cleaner and glossier, a sharp contrast between the blue-black of the back and crisp white of the rump and belly. Just as the speed of movement is too great to count an overall number, so I cannot discern a ratio of adults and young, but I can say with some certainty that there is nothing other than house martins in the flock. A good number of birds seem to be taking an interest in a lone hawthorn that stands some 80 yards or so along the fence line. It is a fair-size tree (for a hawthorn), particularly given its position in the buff of the prevailing wind. The branches on one side (facing west) curl back upon themselves but are also slightly stretched, the bounce of the wind against the ridge, which provides such irresistible lift for raptors and gulls, also offering the tree an upward boost. The martins could be feeding upon ants. I have puzzled before when seeing martins and swallows massing in the spindly ash tree just below our garden, before realising that a huge hatch of ants was climbing the limbs and using it as a launch pad – the birds saving energy and picking them off the branches rather than catching them in the air. Given how indifferent the martins seem to be towards me, I will inch along and take a closer look.

Bird migration has long puzzled us. Species such as the house martin and swallow, which are so conspicuous in spring, particularly when making use of man-made structures to nest, would simply vanish in the autumn. The earliest known English treatise on the subject came from Charles Morton in the late seventeenth century (there seems to be some uncertainty as to the exact year). Morton was an educator and Nonconformist minister from Cornwall whose beliefs and progressive teachings led to his arrest and excommunication, and a subsequent relocation to the United States where he took a role at Harvard University. He theorised that swallows did not hibernate, as was the popular belief, but spent the winter on the moon. The journey (flying at around 125 miles per hour) took two months in each direction, leaving them four months to breed and feed on earth and an equal time surviving on fat reserves in the lunar landscape.

Morton wrote that the absence of birds in autumn 'is such that we know not whither they go, or whence they come, but are, as it were, miraculously dropped down from heaven upon us'. Words that stir thoughts of the painted ladies that I watched dropping from the sky in 2009. Had I not known otherwise, through the proven study of others, then it would not have crossed my mind that they might have flown all the way from North Africa. So just as Morton's thoughts might cause a smile today, he wasn't too far away from the reality. He was certainly a step closer to the truth than the notion of swallows spending the winter in the mud of lake beds or tucked asleep in crevices of tree trunks. And, similarly to Gilbert White, he was writing without resource and to a limited audience.

When we ponder migration so we often tend towards those species that travel great distance. The hirundines and warblers, or antipodal-travelling Arctic terns. But movement of some sort occurs across the avian world, even if it is a matter of yards. The robin that nests in the garden hedge is likely different to the one that is so possessive of the bird table in winter, but they may have spent the counter season in one another's place. For three winters I was well trained by a robin that appeared whenever I arrived home in the car. It knew that food would always appear as I opened the door (it was particularly partial to Bombay mix) and so would wait expectantly for it, even coming into the car itself if I didn't dish out dinner quickly enough. The bird vanished that first spring, but reappeared in autumn, before, come the following April, taking up territory in our garden hedge. It had to be the same bird – the mannerisms and confidence were too familiar for it not to be. And one day, that following autumn, I wasn't greeted as I stepped out on to the patio but found a familiar face a minute later as I unlocked the car.

The cloud of house martins will be travelling a lot further than car-park robin ever did, although I cannot be sure whether they will take the flight further south before it is dark. They will travel at night-time but might also look to find a roost site and it seems as though that might be the cause of the interest in the hawthorn tree. I have inched as close as I dare and the martins still appear untroubled by my presence, but I have no wish to spook them and will edge no closer. They do not appear to be feeding on the tree, but are dropping down to line the branches before lifting off and rejoining the mass of low-flying hawkers. It seems that they are testing out this spot as a potential overnight roost. There is obviously plenty of airborne food to keep them busy until darkness, after which the hawthorn will provide a perch. Perhaps the adult birds are landing to demonstrate the plan to the newly fledged youngsters, one advantage of travelling in mixed age groups being that the older birds, who have completed the journey before, can share successful strategy.

There is so much movement and interaction that my binoculars have been something of a hindrance. It is almost impossible to follow an individual bird without my eyes being whisked off by the movement of another. It is like picking out a single snowflake in a wind-driven blizzard. But I am able to study the birds in the tree far more easily. The sleek lines, slim body and the round, black bead of an eye. The beak is quite small but belies a wide gape, and the feature I notice most is the feet. They are delicate and crisp white, gloved like the hands of a snooker referee but with a light, slightly fluffy, feathering. I have read that this is to offer warmth while roosting on the wing, although that might contradict the purpose for which I presume they are visiting the tree. Perhaps, then, they are actually taking the chance to rest before overnight passage at altitude.

That so much mystery surrounds such a familiar animal is both a fascination and a concern. The house martin is a bird in decline and, in order to find out precisely why, we will need greater knowledge of their lives. Lack of food is certainly an issue, and distribution maps produced by the British Trust for Ornithology point to a huge decline in England but less so elsewhere in the British Isles, with the populations actually increasing in parts of Scotland and Ireland. This does suggest a link to the lack of insect life due to intensive farming, but other factors must also be considered. The increase of extreme weather that downed so many birds last autumn can also be felt in spring. This April was the sunniest ever recorded in Britain and also one of the driest. For birds reliant upon mud for nest building this will have presented a major obstacle. The martins that have long nested on the houses in this village have always favoured a puddle at the end of the lane, which forms at the bottom of the hill and takes time to drain. In the last few years, however, that puddle has dried up at a time when the birds have needed it the most. I began walking up there with buckets filled from the water butt and other villagers began doing the same. One of the local farmers would use the front-loader bucket on a tractor to transfer a huge scoop from his pond, but for all of our efforts, the local population has tumbled. There were at least two dozen pairs nesting when we first came here, but last year there were just six and this year only two. The people who live in one house spoke of the numbers thirty years ago, when there were nests along every eave. They would put wire about the doorway so as to limit the amount of martin-poop that came into the house, but this year they had not a single pair.

It is a sad shift. The house martins have been part of the identity of this place; a constant presence through the summer months, a coming and going and the build of numbers as the fledglings join the buzz and chatter of the evening's sweep. But right now, rather than dwell upon that disappearance, I shall enjoy the spectacle of today's appearance. This may be a fleeting visit, but it is also an opportunity for me to absorb and appreciate the moment. I will withdraw from the hawthorn tree, though. Just in case they are looking to stay for the night.

The sky has been washed with a thin layer of mauve. The cloud is thin but substantial enough to water down the sun which is beginning to melt into the evening. It remains above my horizon, but our cottage below me is already in shadow. To the north is a more substantial billow of cloud, but not enough to threaten any rain, and my eyes are drawn to the metal masts on Rampisham Down. Only three remain, from twenty-six that stood a decade ago. The tallest masts measured more than 300 feet and the site was one of the main transmitters of the BBC World Service. At first, I rather resented seeing them on the landscape, a stain of metal in the thick of the rolls and wilds. But in time I came to appreciate them, just as I do the pylons that march to the south, not least because they provide a platform for peregrine falcons and ravens to nest. The downland at Rampisham is special too. A cap of acid grassland that sits atop a chalk mass, allowing a mixture of plants and fungi with completely different pH dependencies.

It is always interesting to see a familiar landscape from a slightly different perspective. I remain within sight of home and in the same small field, but have never before stood in this spot. I look for other familiar landmarks, the masts to the east and the small, single oak tree that stands beside the road on the next ridge to the north. The oak takes a bit of finding, but just as I lower my binoculars a yellowhammer lands on the hedge-top around 20 feet from me. It eyes me warily, a slight flick of the tail that causes it to slightly rock upon its perch. As I watch, a movement below catches my ears and eyes – something considerably more substantial in size than a bunting. Rabbit is the obvious source, but it sounds as though it is awkwardly working itself out of the hedge on the other side and a rabbit would likely either sit stock-still or bolt underground. It could be a fox – there is a den nearby – but I am rather hoping it is a hare. We don't see many brown hares locally and a glimpse is always a treat. They can appear languid when idling along the verge or field edge, or otherworldly when tucked up in a form with long ears flattened beneath those extraordinary amber eyes, large black pupils and a dark ring around the edge. But when they take flight they move like the wind, powered by kangaroo kicks from those gangly hind legs. That speed might point to whatever was scrabbling beneath the hedge, because in the moments that it has taken me to take the few paces to the source, the culprit has gone.

I follow the slow curve of the field edge until I reach the corner, now able to look back at the hawthorn from the opposite direction. The martins are still busy, but more birds seem to be perched now. Perhaps they are tucking up for the night. Before me is the combe that funnels the western breeze, bumping into the ridge I stand on and throwing up those lovely updrafts that birds love to hang on. I've missed the sun now, a slight shame because it casts some lovely long shadows across that piece of valley. From our cottage, the shadows of the trees on the western edge of the combe stretch all the way across to the treeline on the eastern side. It only happens for a few weeks either side of the spring and autumn equinoxes, when the sun sets at that specific angle, and it would have been interesting to see them reach towards me. The haze would have diffused their impression, though, and the day is drifting into one of those chilly autumn evenings when people might light their fires for the first time since winter – smoke curling gently into the pink. I rather like the autumn coolness. Outside, the air feels clean and liquid, the grass thickening with dew. Inside, though, a dryness remains from the heat of the day with the promise of more tomorrow. It's time for evening jumpers but we'll go without a fire for a while yet. And the still isn't limited to the temperature and the thinness of the air. The world is quiet. Not eerie, but watchful. As if waiting for the change. As the house martins head south so the first of the winter visitors will be arriving from the north, while the songbirds that remain are tucked out of sight, fledglings flying their own path while their parents moult into winter clothes.

Hah! As if on cue, the hush is broken by the needy wail of a young buzzard – not all of this season's youngsters are cutting their own path. There was only one successful buzzard pair in the village this year, but they did manage to fledge two chicks. Then, from mid-June when they first took flight, the desperate cries for food filled the soundscape from dawn until dusk. The call is not the evocative, almost mournful mew of an adult, but a 'Muuuum, Daaaad!' wail that is not particularly useful for a bird trying to sneak up on prey. In recent weeks, only one of the two has been calling. Hopefully its sibling has learned the art of stealth and is finding food for itself, but it might well have perished, living as they do in that perilously high seat, dependent upon everything beneath them functioning as it should. Their number locally seems to have correlated with the fortunes of the rabbit, a prey species the buzzard is highly reliant upon. When we moved here, I would surmise that the buzzard population in the folds of West Dorset was close to optimum. There can never be too many top predators, of course, but it appeared that their number was thriving. In our first winter, as the farmer pulled the plough across the field above the western combe, I counted twenty-seven buzzards on a single thermal. They would frequent telegraph poles and fence posts, and soar over every hillside on a sunny day. But a couple of very wet summers saw poor recruitment and, although good numbers remain, they are not as plentiful as they once were. A few familiar faces have gone too. Buzzards vary enormously in plumage, from an almost complete covering of chestnut brown to a pale, almost ghostly white. This can make individuals easy to identify and that then leads one into the dangers of oversentimentality due to overfamiliarity, although I am not concerned for the long-term fortunes of the species. It will always be my favourite bird, a legacy of childhood holidays to those remote, westerly places where buzzards had found their range restricted as a result of persecution and pesticide use. Their revival and nationwide spread have been dramatic and their adaptability to this modern environment the main reason why they should endure. Away from the realms of eagles they are the bully of the bird world. Easily outmanoeuvred by peregrines or red kites in the air, but the boss of them on the ground. They would come off second best in a tangle with a goshawk, but away from the forests the buzzard should be able to pinch food from anyone else not strong enough to carry it away for consumption. And they are none too fussy about the source of their protein either. Earthworms are a favoured food, hence the cloud above the plough, while beetles, frogs, crickets and grasshoppers will fill that proverbial hole. It could be that the buzzard has filled the niche that the kestrel has sadly vacated, but we should not blame it for that. Habitat, as always, is the key to a species' survival but we tend towards fixing the numbers rather than providing what is needed.

I pause between slightly gingerly steps as I ease down the slope, looking up at the changing sky. There is very little red in the sunset, but mauves and a slight blush of pink. The blue overhead is rapidly deepening to navy and I glance around wondering if Venus has begun her glow. And then I hear it – singing. Of course! They are not just buzzard food but the voices of *Cantlos*. As the birds quieten so the crickets sing. I am not going to pretend I know which species I can hear, and, in truth, I presume they are crickets rather than grasshoppers because they are, typically, crepuscular while grasshoppers are diurnal. But much as people can enjoy the avian dawn chorus without knowing who is singing which line, so I can stand for a moment and enjoy this cacophony. I could only name one or two species of grasshopper and cricket with any confidence and that would be through sight. The sounds I have never even attempted to isolate, and it is interesting how difficult I am finding it to pick out individuals. I hear one close behind me, a tentative rub of the back legs – a stridulation – as though aware of me and wary that I might spin around in response. And, of course, the moment I do it falls silent and I see nothing, just the thick rolls of green and yellowing grasses in the fading light. I think I pinpoint another, a few paces to my left, but despite the greater vigour, the sound rises and falls, presumably due to the insect moving in order to cast a greater range. A move that may draw the attention of a female but with the added bonus of confusing potential predators. I am not going to successfully discern one species from another, but I do enjoy the chorus. It mingles with the sweet smell of the dew to register somewhere in my psyche as a trigger combination that signifies early autumn. I must hurry back home, though, and ready myself for the moonrise. I cannot remember the exact time it is due, but it could already be nudging into view were I still up by the gate and I want to watch it from the comfort of home.

There is an undeniable logic behind the creation of a lunar calendar. The shortfall in terms of time when correlated with the solar year is obviously an issue, and something we have already considered, hence the apparent sixty-month cycle demonstrated on the Coligny calendar. But what makes perfect sense is the use of the moon as a medium. It might come and go like the sun and be masked by even light cloud, but it is consistent. And the moon that is seen above Eggardon is the same moon that rises above Lewesdon Hill, Danebury Hill and Santa Tecla in north-west Spain. The Celts could make arrangements, for trade or moot meets, and know that all parties would remain chronologically aligned.

What would they have thought of the moon itself, though? How might they have reasoned its being?

As I watch it climb above the ridge, breaking the milky shackles of that thin cloud, I cannot help but be in awe. The moon is almost full, squeezed slightly on the left-hand side as it waxes into its whole. And the cloud seems to trail from it like a silken veil, the glow igniting the wisps that then shrivel and fall into the black. It took an age to appear, the yellow-tinted loom of its coming seeming to stall in a moment, until there it was, already well clear of the skyline but sneaking a secret path behind the cloud.

As with so much of Celtic culture, little is known of their deific beliefs. Barry Cunliffe points to Julius Caesar who touched upon the subject in *Commentarii de Bello Gallico*. Caesar wrote of the many Gaulish superstitions and attempted to juxtapose Celtic gods with those familiar in Roman culture. It seems he was attempting to simplify an understanding but from a set perspective. Much as I have done when spending three-quarters of this year wedging a variety of Celtic polygons into my nice, neat circle. What seems likely, though, is that the Celts associated spirits or totems with specific objects or phenomena rather than personification. That said, there are legends existing within Irish and Welsh traditions that have human basis and vague links to the moon. Names such as Cerridwen and Arianrhod in Welsh mythology and Morrígan in Irish, but they have no definitive roots within pre-Roman society.

The moon does feature extensively throughout documented ancient culture. The Romans and Greeks looked to Diana and Artemis, although both were goddesses with influence that stretched beyond the lunar. Instead, we might look to Luna and Selene, who, in their respective Roman and Grecian persuasions, had a more direct connection – both driving chariots across the sky, carrying the moon as they go. Given the Celtic reverence of horses and the value placed upon riding and chariot driving, a similar figure may have existed. If a chieftain was deemed worthy enough to be buried with a chariot, presumably so as to make safe passage into the next world, then perhaps the moon might have been granted similar transport. This may well have been the case in later periods of Celtic history, particularly as their cultures began to mingle with those of Greece and Rome, but before then, around the time that the fortifications on Eggardon hill fort were first created, the use of horse-drawn vehicles was unlikely to have been an influence if used at all. Instead, perhaps there was a similar figure to that of Jie Lin, worshipped in ancient China, who carried the moon in his arms. Or Khonsu, a god of ancient Egypt, whose name likely translates as 'traveller'. He had a strong association with both the moon itself and the passing of time, controlling both lunar movement and cycle. Such influence reached deep into day-to-day life, prompting bouts of insanity (linking with 'lunacy' in more contemporary Western history), and even death. When taking the form of a crescent, Khonsu would ensure the fertility of people and livestock, helping plants to germinate and fruit to ripen. There are definite links within such associations as those suggested by Pliny in relation to the Gauls. Not least within the peculiarities applied to time and the importance of completing tasks at specific points within the lunar cycle.

Time recurs as a lunar theme across many cultures, and in Norse mythology it was Máni who was responsible for its passing. He was god of the moon and brother of Sól, goddess of the sun, both spending their lives pursued across the sky by giant wolves. It was said that when the wolf drew close to Máni, so it would begin to mask the moon, causing an eclipse. And though Máni would escape his fate, he would not manage to forever. The wolf was destined to one day catch and devour him, plunging the world into the turmoil of Ragnarök, the passing of all of the gods and the rebirth of a new fertile world.

References to these legends were found in scriptures discovered in 1841 by Georg Waitz in Fulda, central Germany. Among the manuscripts were two spells written in Old High German and dating from the ninth century. These are known as the Merseburg charms and are believed to be incantations recited to a particular end. The first is a blessing of release, while the second charm is recounted to cure a wounded horse. It mentions Wodan, who was the Germanic incarnation of the Norse Odin, and Sunna, an alternative name for Máni's sister Sól. These charms were surely not isolated spells, the specific nature of them suggesting that there may have been many covering a range of needs and occurrences. And if such incantations were familiar within Norse and German societies, then it makes sense that similar chants and songs may have been performed a little further west.

I'm trying to wedge a giant rhombus into my imaginary circle here, but as I look at the moon now, so big, assuring and dependable, so I might feel a compulsion to chant in its honour. Particularly if I were to find myself on Eggardon, 3,000 years ago, a familiar murmured incantation coming from those around me. A rhythmic, almost hypnotic and ultimately irresistible seduction of sound. Just as I might unconsciously tap a foot to a favourite song. The Moon of Singing. Of incantation. The key now is its prominence.

We spend the summer giving little thought to the moon. The night sky comes late and is slow to darken, while the pale of the day, below the bright, relentless sun, will hide the white unless we search hard for it. But as the equinox approaches so it booms back behind a short dusk and into a blackened sky. It is almost insistent. And what if, before a fire is lit and while the air is not too cold, people took this opportunity to sit and watch the moon for no other reason than appreciation? The first ale might have been brewed, *Amanita muscaria* gathered from the woods. Not only would the moon glow, it might smile and wave, the constellations dancing around it. That would be enough to get everyone on their feet, singing and dancing.

I think I'll stick to a glass or two of cider tonight, though. It is Saturday. But we'll leave the television off for a little while yet.

10

THE HARVEST MOON

Stirred for a bird, – the achieve of, the mastery of the thing!

(Gerard Manley Hopkins, 'The Windhover')

I VISITED STOURHEAD a few years ago while researching another book. I wanted to see the source of the Dorset Stour, a river with which I had become increasingly familiar. I wasn't making some affectional pilgrimage, though; instead I wanted to ponder the meaning of beginning. I had begun to see the river – any river – as a metaphor for life, in the way that it meandered as a small stream, then built power and directness as it grew, collecting tributaries and overcoming natural and man-made obstacles, before fading in identity and losing itself to the salt of the sea. Its source, though, was indeterminable. Does a river begin at the point where it first springs above ground, or in the underground reservoir where its waters gather? Perhaps it goes back further, to the point at which a raindrop lands or a cloud forms, or later, to a place where the flow is discernible and untouched in drought.

The creation of a human life is equally contentious. We celebrate our birth, but have already experienced nine months of growth, so where do we actually begin? At the moment that our brains begin to mould, or should we consider the point of conception, or the conception of our mothers and fathers? We remain dependent when the umbilical cord is cut, wholly reliant upon our parents for food, hydration and warmth, but at what point do our personalities form? How much of what makes us is decided long before our physical being?

I found the complexities fascinating and, without wishing to dwell on a subject already explored, it is a theme familiar wherever we look. A mayfly spends the briefest, ephemeral part of its life cycle above the surface of the water, yet we know it as a mayfly. A butterfly might spend longer in the form of an egg, caterpillar or chrysalis than it does an adult, so at what point does its life begin or does metamorphosis mean it lives more than once?

Then we might consider time and its passing. The stiff structure that we place upon it. A 'year' is defined as the length of an orbital period of a planetary body, exclusive and specific, whereas a 'day' is the length of time taken for a planet to complete a single rotation of its axis. We fit 365 days into a year with the leftover quarter collated and added every fourth year. And it works very well across different languages and cultures, although there is a curious and rather unfounded rigidity to a year's beginning and end. A solar or tropical year can only truly be measured by the completion of seasons between two consecutive equinoxes, and 1 January, although just ten days on from the winter solstice, is perhaps not so significant as we make it out to be.

I am frying my brain with such thoughts, and if I'm not careful of my footing I might tumble into a realm of quantum physics that I really don't have the cognitive ability to make sense of. And to be honest, I have brought this upon myself. I was going to start this chapter with a cheery 'Happy New Year!' before realising the dubiety of such a statement. And, in truth, I have rather lost track of the days of late – and of the moon. We have certainly moved into the next Celtic 'month', but the question of whether we have entered *Samonios* is complicated further by whether a 'year', as we understand it, was recognised in a calendar that apparently functioned across a period five times as long.

Happy new something, then. And perhaps the ambiguity is apt. The 'new year' as we recognise it has long left me cold. I do not mind the festive period and have never been one to turn down an opportunity to overindulge into the early hours (although that is becoming less of a pleasure with increasing age), but the notion of a new start leaves me cold. Aside from the heavy head, the first day of January feels no different to any other day. The sparrows still squabble over the peanut feeder and the robin still stands expectantly on the rim of his (or her) special flowerpot. I fail to feel the need for resolution or fresh beginning, of fasting or new-fangled methods of self-deprivation. I find it tough enough to be nice to myself without inflicting a self-prescribed regime in the darkest of days.

Gosh, I must sound grumpy – and I'm not. I am, however, beginning to feel the pinch a little as the days shorten and I definitely need to redden my cheeks and build up a bit of a blow. But I am just about to boot my feet and do exactly that. A walk that will begin from somewhere new despite it being only a couple of miles from home. This is one of those spots that fall upon the grey edge of a triangle. I know the other two sides well – they converge close to home and lead respectively to all points north and east. But the link between those routes is a side I never need to use. Which is why, despite living here for ten years, I have never before opened this gate and followed this track to wherever it leads. I have a pretty good idea, of course; the footpaths are marked on the Ordnance Survey map that is folded in my shoulder bag. But there is an excitement of adventure and discovery – and options. If footpaths cross then I'll walk where I feel compelled.

I'm not religious, but neither would I call myself an atheist. In my mind, an atheist is someone who actively disbelieves and that isn't me. Belief is one of those threads of the unknown that can never be proven or disproven, and those invested, on either side of that cause, will never be wholly convinced of the alternative. So rather than worry myself over arguments or counterarguments, I tend towards the subjects that are either resolvable or dressed with a little predetermination. The mindset that I applied to my understanding of the Coligny calendar.

My brother, Rich, once laughed and asked me how I could not believe in ghosts when I had seen one. It was a fair point, but much as I know what I saw (in the instance to which he was referring), I cannot wholly explain it. A 'ghost', in terms of popular belief, is a thing that I am unable to align with my knowledge of the world. I don't accept the idea of a soul existing after death, of being able to manifest or haunt or interact. But equally, just as time is a dimension that cannot be boxed to suit our needs, so the residues and boundaries of energy and molecular structure are not as clean as we tend to perceive. The majority of us like simple, straightforward explanations, a result of taking a singular stand and closing our ears to any alternative. Something that is drilled into us from an early age. We are exposed to structure and rigidity from birth, particularly in the West, and even if our immediate situation is chaotic then it will be shackled by society. Rules can be hugely beneficial, of course; they help us co-exist and create a moral balance, but they also restrict individual thought and convince us that being different, living outside of the accepted norms, is wrong.

But what then happens to that endless curiosity of childhood? The fascination with *everything*, the desperation to be overwhelmed with knowledge. Often, it remains suppressed, but should it re-emerge, sparked by trauma, illness or heartache, then it is at a point when our minds are vulnerable. The tap that we switched off in childhood is not gradually tweaked to release any of the pressure but wrenched wide. A sedated mind suddenly open but without the protection of learned cynicism. It is a conspiracy theorist's dream target, an adult asking questions but hearing the answers with the innocence of a child.

Anyway, I digress slightly. The hypothetical vulnerable mind to which I am referring was, of course, my own. And I am grateful now for having been emotionally exposed while relatively young. Accepting my own insignificance was the liberation my troubled psyche craved. That may sound like a contradiction, particularly when considering the very dark corners into which I slipped, but in understanding my own ego, I realised that I did have choices. Those feelings of being special or unique, and troubled to the extent that I was, came about because my being felt too consequential for me to be anything but. They were the same protections that we apply to an afterlife, or heaven and hell, or ghosts. How can my self, this mind, actually stop being? It is an extraordinary arrogance when you stop and think about it, but just as it is important to accept one's own insignificance it is vital to remain open-minded. It is not to simply accept the inevitability of finality but instead understand that there will not always be answers, that it is okay to be wrong, to change an opinion, to not know everything. It might not have been a ghost that I once saw, but I did see something. And that something cannot be explained in terms of accepted norms – it cannot be neatly packaged and filed. Perhaps it was a subconscious feeling that I put physical form to or an energy that had stained the ether. And there is much to be said for a sense of feeling in a place. Energy is residual. The sanitised landscape I visited with my family at Bosworth may have felt empty simply because it was mislocated, because the connections I feel up on Eggardon are familiar in other places too. The atmosphere of a church is invariably one of peace and comfort, a reflection of the energies that have been left there. I do not need to believe in God to feel the positivity and sanctuary of a building built and used as a place of worship. Conversely, a landscape or place completely innocuous in appearance can feel unwelcoming or even threatening. If a trauma leaves a vestigial mark then perhaps our minds, when sensing it, create their own manifestation in response.

And these ponderings are pertinent because, as I walk, this pathway feels like the dividing line between two very different landscapes. To my right is pasture, rolling down into the outside of a horseshoe-shaped combe. There are sheep grazing the field closest to me, but the folds are lush and inviting. A hedgerow edges the path to my left, impeding my view of the shallow rise behind but also protecting me from it. I pause beside a gap in the hedge, formed apparently by the force of animals – almost certainly deer – and spend a few moments looking. The ground is stubbled, the relative flat offering the opportunity for arable cultivation. And at first glance it seems busy. There is movement among the yellowed stalks, but a scan with binoculars reveals nothing other than pheasants and red-legged partridges, likely releases for this autumn's season. At the back of the field, to the right-hand side and straddling the brow, is a conifer plantation, and while trees of any sort will normally add a depth of balance to a view, these seem cold and out of place. There are not enough to offer an identity of their own, and instead the green looks strangled by the browns and yellows in the flat of the surrounding field. The components of the view might not be especially inspiring, but they do not warrant the depth of discomfort that I sense. Perhaps it is the way that the field vanishes over the horizon; I am troubled by not being able to see the far hedge line and what lies beyond. Whatever the reason, I have no desire to continue on this current course. About 50 yards ahead is a right of way across the fields to my right and I will gladly follow it.

It is interesting how a sound, initially so intrusive, can become so familiar that it all but vanishes. For many years we lived beside a railway line, with trains thundering past on their way to and from London Waterloo. Visitors would often remark upon the noise and vibration, but we barely noticed it, unless the rattle came at a pivotal plot point in a film we were watching. Odd too is the fact that we might then not notice when the sound ceases. Like with a niggling cough, we are aware that it has gone, but do not recall the precise moment that the tickle stopped.

On the edge of our village is a grain store and dryer, and such has been the need this harvest, the fans have been whirring almost constantly. After that cold, dry spring, the wheat and barley fields clung on to their green, only to be dealt the sudden scorch of late July. But they were dried out rather than ripened and the combine harvesters have had to cut through the unsettled late summer and heavy dew of early autumn. Crop that was already short on yield and quality would require the added cost of drying to prevent rot. This pattern was not echoed everywhere in the British Isles, however, with recently released preliminary figures from the Department for the Environment, Food and Rural Affairs pointing to over 14 million tonnes of wheat produced in the UK. This is a marked increase upon 2020 and the yield itself, at around 7.8 tonnes per hectare, is close to the long-term average. The record yield for 2021 (and also the record for Britain) came from a farm near Louth in Lincolnshire which produced 14.31 tonnes per hectare. To place these figures in context, in 1701 it was estimated that the average wheat yield was 16.2 bushels per acre, which, if my mathematical knowledge hasn't completely left me, would equate to a little over 1 tonne per hectare. An eight-fold increase in the average yield shows the remarkable impact of farming development, yet farmers today might still be cutting at a loss. The lines are so fine, demand so great, that further advancements will always be sought. Yet to do so, we are effectively sterilising large swathes of farmland. To create more organic, ethical farming opportunities, we must work from the point of demand and ensure the farmer is paid a fair and consistent rate.

Another thought on the figure from 1701 is how much that will have increased from the early days of farming. When wheat was first cultivated by humankind, some 10,000 years ago, the plant itself would have looked very different to the ones we see today, requiring more space to root and growing as tall as the people who cut it. From those early beginnings in the Fertile Crescent in the Middle East, farming was surprisingly slow to spread. It took nearly 5,000 years before methods of crop cultivation were adopted in the British Isles, around the time that the Celts were constructing the first hill forts. Those first beginnings must have been tough and very much a case of trial and error, and the harvest itself would have taken longer. With no machinery and only rudimentary tools, the work would have been laborious and for negligible end product. Bring down an aurochs or deer and there is food for all but fill a bowl with grain and your fellow tribespeople would probably be less impressed.

It makes sense, though, that the Celtic 'harvest' would have continued until summer's end – *Samonios* – and the feast of Samhain. With winter posing such a threat to life, particularly further north, it would be vital to ensure everything was in order before its coming. Early Irish literature points to a time when cattle were rounded from the pasture and some slaughtered for food, and also when doors opened into the Otherworld, allowing spirits and deities to move among men. The *Aos Sí*, 'the people from the mounds', the souls of the dead or fallen deities, would cross into this world during Samhain and offerings were made to appease them. Fires would be lit, often on hilltops such as Eggardon, the flames possibly symbolising the sun and a blessing and thanks for the heat and life it provides. The fires were linked to divination rituals, possibly precursors to human sacrifice, with some activities (sacrifice hopefully excepted) recurring through time and culture. Indeed, the date at which Samhain is recognised, 31 October into 1 November, is marked in contemporary society as All Hallows or Halloween – although the original motifs and practices have been replaced by door-to-door collection of sweets by children in fancy dress and rows of moulded plastic lining supermarket shelves.

Amid the celebrations, the Celts placed a strong emphasis upon fertility. With the produce of this season gathered, it was now time to ensure that the following year would be a success. As spirits moved freely between worlds, the god Dagda would meet with the goddess Morrígan, and their sexual union would ensure the wellbeing of the people. Such values and beliefs point strongly to the closing of one season, the end of summer and of harvest. All that could be taken from the soil and land had been gathered and Samhain was an opportunity to give thanks and ensure future prolificity. It makes sense therefore, to me at least, that the full moon rising in a little over forty-eight hours' time should be called the Harvest Moon. Contentious because of the association with the autumn equinox, as I previously mentioned, but also because this imminent full moon is similarly notorious. In fact, as I become increasingly aware of my own contradiction with common belief, so it seems that public interest is rapidly increasing. Talk is already rife of the 'Hunter's Moon' that is waxing close to full. Again, the timing of its rise, coinciding as it does at this point of the year with the setting of the sun and the blackening sky, cements the curiosity and fascination. But I have no intention of bowing to North American custom – bad enough that it is partly the influence of modern America that has filled the shelves and landfill sites with plastic tat. The Harvest Moon it is.

'Kek-kek-kek-kek!'

I stop still and turn. It is faint and fades slightly on the wind. A kestrel. I cannot see it; the grassland wedge that I have skirted alongside and now dropped beneath is blocking my view. Briefly, though, the breeze eases and the sound soars against it. Short shouts not unlike a green woodpecker's in rhythm and tone but more consistent and sharp-edged.

Our four native falcons have similar calls, yet each is distinct. The peregrine is the deepest, with each note held the longest. It carries for some distance, bouncing off rock faces and sea cliffs, a critical part of the soundtrack to the wild places. The merlin sounds like a hurried peregrine, a similar rawness, almost an edge of fear and panic, to its voice. A hobby, like the merlin, is a sound I had not heard until quite recently, but I stumbled upon a nest site where the youngsters were just fledging and the noise was almost constant. The sound is the most polished of the four, clean and sharp, much like a hobby is in flight.

As with the young buzzards at home, falcons are most vocal when still dependent. So perhaps the kestrel is a juvenile bird in one-sided conversation with its parents. It has determined the next part of my route at least, and the paths I am walking today are delightfully vague. The sign that earlier offered me escape from the original track was pointing into an open field. It was empty, but well grazed and, judging by the freshness of the pats, very recently vacated by cattle. I followed the fence line and found the next footpath sign, and stile, a third of the way along the southern edge. And crossing at that point took me into the landscape that I had felt invisibly drawn to. Unimproved grassland, tufted and lumped with anthills. There is a scatter of sheep in the next field over, but the pasture appears only lightly grazed. The difference in colour is marked, from the lush, deep green of the field I crossed to the paler lime and coarse, fibrous growth around my feet. A closer look reveals a multitude of species and to the touch of my hand it feels substantial, almost pillow-like.

To my right, a shallow cleft is gathering width and depth before disappearing into the main combe that runs in front and beneath me, perpendicular to my position. To my left juts a wedge that blunts and drops quite sharply at its end. It was from the other side of this wedge that the kestrel called, but I shall work my way around the front rather than tackle it head-on. That way, I am keeping a vague (and excusable) course that matches the rights of way marked on my map, while also offering views up and down the body of the main combe. The wind is also less keen on the lower slopes; it is south-westerly, and mild, but it carries a threat of rain. I did put on a waterproof coat which, though weighty and slightly suffocating, might prove useful before this walk is completed. For now, though, I shall enjoy the broken sunshine and crackle of light.

I climb the next gate, an old, slightly rusted seven-bar barrier that still looks considerably more robust than the single-step stile next to it. I am now among the sheep and try to move slowly as to avoid spooking them. It seems though that I am a figure sparking curiosity rather than fear, and although a few wander off, most eye me with interest as they continue to chew the cud.

'Kek-kek-kek-kek!'

It comes from behind my right shoulder this time and I turn to find not one but two kestrels within 50 yards. The first is in a hover, just over the edge of the main lip of the wedge. In fact, it's a hang rather than a hover, the breeze pushing up perfectly for it to sit without effort. It looks almost glued in place, while the second bird, the one making the noise, swings around behind it. It is, I presume, a young bird pestering a parent, but it seems to be learning for itself because it quietens to the indifference it receives and then fixes on a hover of its own. I rest my bag on the ground and kneel, first checking quickly for sheep-shit, before wedging my right elbow against the slope to create a decent stabiliser for my binoculars. But in the moment I put them to my eyes I am distracted by movement almost immediately above me. A third kestrel, this one no more than 30 feet from me. Again, it hangs rather than hovers, and I try to watch it without moving my head. A sneaky sideways peek to pretend I haven't actually noticed it is there at all, and it seems content. My binoculars would be a hindrance at this range. I briefly think of reaching for my camera – I will struggle to get another close-range opportunity such as this – but for the very same reason, I want to enjoy it with my eyes.

The head is absolutely still, its gaze seemingly fixed on something further up the slope. It is close enough for me to make out the yellow of its eye, and, as I watch, the body of the bird seems to tilt forward very slightly. The head remains still, but now the wings have angled into a more familiar pose, with the occasional flap to hold station. The tail fascinates me. Often, a kestrel will fan out its tail while hovering, but this bird is holding its own shut. The side feathers seem to flare slightly, and the whole tail twists like a rudder in response to the wind, but either this bird is still learning the art or the uplift from the breeze is too strong for it to completely open up against.

While the poetry of T.S. Eliot left me slightly cold at school (at this point, having mentioned rather critically 'The Love Song of J. Alfred Prufrock', I decided I should reread it – and I rather like it now), the words of Gerard Manley Hopkins clobbered me square between the eyes. Unsurprising, I suppose, that 'The Windhover', given its subject matter, would capture my imagination, but I have yet to read another poem that opens quite so beautifully:

> I caught this morning morning's minion,
> Kingdom of daylight's dauphin, dapple-dawn-drawn
> Falcon, in his riding.

Those words repeat over and over in my head. So perfectly appropriate. I only wish I could remember the rest of the poem. This moment lasts only a few seconds, but time seems to stretch like elastic. The kestrel has swung away now, a little further along the ridge, and I feel spoilt by the encounter. I have had some incredibly close meetings with wild kestrels before, invariably young birds who seemed fearless of me, but they have always been perched. To have such intimacy with a windhover hovering has sent me into a heady spin. It still works the ridge and still the first two birds quarter further to the west. And now a fourth drifts over, a couple of sharp swings across the thermals before meeting the lift that the others ride. For several more glorious moments I have four kestrels hovering within a short stoop of me.

'Kek-kek-kek-kek!' 'Kek-kek-kek-kek!'

Now there is bickering and movement. Lesson over, perhaps. School's out. One bird remains in quiet but the other three chase and harry one another back over the top of the wedge and out of sight. Their voices disappearing once more on the wind.

Woah! Now I have a fright. Turning before standing I come nose to nose with a sheep. Clearly wondering as to what I was wondering, it edged up for a closer look. The poor sheep jumps more than I do and it flees, causing mild panic among the nearest edge of the flock. The calm returns quickly at least.

I have a head full of kestrels, of the elegant slide and sharp stops. They are a bird I look upon so differently today, so familiar were they in childhood. Aside from glimpses of tawny and barn owls after dark – normally a fleeting flash in car headlights – kestrels were pretty much the sole source of my burgeoning raptor obsession. It wasn't that I didn't enjoy seeing them; it was more that I was desperate to see other species. It is sad then that the kestrel's demise – especially across the areas where they were so common forty years ago – has in part prompted my appreciation of them today. I like to think that I see the world slightly differently, and am able to find beauty in the familiar, but sightings are definitely loaded with added emotional weight. Much as I love seeing them in flight, though, my favourite sight is of a bird perched in its kestrel 'hunch'. They are normally watching the ground at that moment, a hopeful wait for something small and scurrying, but they do so in distinctive and endearing poise. A head sunk into the puff of the breast feathers, one talon sometimes lifted and limp. A wizened grandmaster pondering their endgame.

As I round the front of the wedge, a narrow but deeper cleft opens up before me. Steep-sided like a gorge, and precise as though sliced out with a giant knife. A loose line of trees runs along the top of the opposite side and among them is a bare skeleton with a very visible nest tucked into one of the forks of the upper limbs. It was likely built by crows, collecting sticks and twigs for the cause, but judging by the two kestrels sitting on the branch beside it, the nest had new occupants in the early summer. Kestrels will often nest a little later than other bird species. A tactic that will enable them to make use of someone else's work. They are not too fussy about the location, nesting on rock faces, building recesses, even using disused rabbit burrows. An old crow nest is perfect, though: decent size and normally positioned high in a tree and with a good outlook to watch for predators.

This nest site is almost unreachable from this angle. I would have to scrabble up the slope to get anywhere close to the tree and there would be no element of surprise in that. Not that I have any intention of doing so, of course; I am happy just to know that these kestrels are here. My route will instead take me on a slow curve away from those trees as it negates the slope on the other side of the gorge. Even with the zigzag, though, the climb is steep. Time to put thoughts to one side and one foot in front of the other.

'Whoosh!' A flash of movement, a startled stumble and a bolt of adrenalin. Excitement, surprise but also guilt. As the rise began to flatten, I drifted off the path, lured by a clutch of puffballs. Lovely round dollops of creamy pink, the skins dotted with tiny white warts that look as though they have been painstakingly placed in pattern. They are small, nothing like so impressive as their giant, football-sized cousins, but always good to find and often leading to other, less conspicuous fungi. As was the case here: on closer inspection, the grass was filled with tiny bonnets and species of the *Panaelous* family that I could not name without a book. With no trees to co-exist with, the mycelia in the meadows break down the roots of deadened grasses and herbaceous plants, often appearing later in the year than woodland species. As I stepped carefully about them, I then had my shock.

A brown hare, launching from beneath my feet as though launched from a catapult. It ran and ran, initially down the slope using its speed to force distance, and then sweeping back up the hillside, kicks and sideways spurts, before vanishing into nothing. Below me, beside my feet, is the hare's form. A hollow in the grass and slight indentation in the ground. It is shaped like a shoe, the grasses at the toe end linked together across the top; the hare must have backed itself in and sat with just its eyes above ground level. It certainly fooled me.

It is said that Boudica, probably the most famous British Celt of all, would release a hare from her gown before battle and let its movement dictate her strategy. Her tribe, the Iceni, famously defied the Roman occupation, Boudica herself fearless and ruthless as she avenged the betrayal of her late husband and violation of her children. The connection with the hare is an interesting one, primarily because it is often suggested that the Romans introduced the brown hare to the British Isles. It seems unlikely that an animal, previously alien, and with such strong association with her enemy, would inspire and guide Boudica's actions. Instead, it might suggest that the blue hare, more familiarly known as the mountain hare, was far more widespread across lowland Britain. A retreat to the uplands, from which it takes its name, having been muscled out by a larger European interloper. And even within an environment that was 75 per cent woodland, it seems unlikely that swathes of pre-Roman Britain would have been lagomorph-free. Recent theory, however, suggests that brown hares arrived in Britain in the Iron Age, somewhere between the fifth and third centuries BC. Archaeological research also suggests that they were not eaten, with skeletal remains showing no sign of butchery before death. Perhaps, the brown hare was not only present but also revered. Animals buried ritualistically and intact. A fact perhaps supported by Julius Caesar's description of the Gauls in *Commentarii de Bello Gallico*: 'They do not regard it lawful to eat the hare, and the cock, and the goose; they, however, breed them for amusement and pleasure.' Proof, perhaps, that 'Boudica's hare' was brown after all.

Uh-oh. A stiffer gust of wind brings a couple of spots of rain. The sun is still shining, but as I turn and look west I see thickening black cloud. It is distant, and seems, for the time being at least, to be skirting to the north. The sunlight feels slightly eerie with such darkness behind it. A moistened glisten that melts on the land like molten gold. It reflects off the steeple tiles of the church roof to the south-west. The small village from where I have explored before. If I had more time, then I would be tempted to head that way. As I walked through the houses last autumn, I saw the charcoal flash of a black redstart, and when I returned I saw it again – this time perched on the guttering of a roof. I had fabulous views, and took a couple of reasonable photographs, and, having seen all of the bird-feeders in front of the windows, decided that the occupants of the house might be interested by what was perched above them. And they were, having overcome their understandable wariness of a stranger clutching a camera and wearing an inane grin tapping on their door.

I press on up the hill, the gradient easier to bear, but the cloud beginning to roll ominously. Pipits and linnets pipe from the barbed wire of the next fence, faces into the wind, bracing themselves for the rain. I'm not sure I'll beat the wet and this route will lead me back to that original track where unpleasantness seemed to lurk. I quicken my pace, stow my binoculars in my shoulder bag and zip them in, and then slip an arm out of my coat to re-shoulder the bag beneath it. I'm about halfway across the field when the first few spots hit, but it's just a splatter – perhaps I've got away with it. Then comes that big roar of wind that precedes a rainstorm and I pause to scan around, looking for options. Towards the far end of this field is a solitary tree – an oak I think – but it does not look very old and the trunk is not thick. By the time I reach it the rain will be on me anyway. My best option is to tuck up and ride it out.

I whip off my coat, tuck my bag up against a tussock and then kneel either side with my back facing the weather. In this position, I can pull the coat over me like a tarpaulin, arms tucked in, hunched like a kestrel. Then the skies open.

For the first minute I am troubled. I tell myself off for dawdling beside the puffballs and overlooking the oncoming cloud. Then comes a brief niggle of self-awareness – what if I am seen, what must I look like? And then comes the reassurance of choice. I could walk on to the car or I could make for the barn up by the main track. Or I could stay right here. This is by far the driest option, and in fact it is almost cosy. And in an instant I am content. I am a hare tucked up in its form, although without threat so I have the luxury of covering my eyes. The rain thumps harder until it rattles into hail. Even better! Raindrops will seep and eventually break through my shell, but hailstones simply bounce off and nestle in the grass around me. I picture the image even if I cannot see it. And I am snug and safe. In fact, I could go to sleep. Right now. I could close my eyes and be asleep before I could count to ten. This is one of those moments where time is mastered. It has no influence. It has no impact upon the decision I made or the events unfolding. I am at the whim of the weather and I do close my eyes – but not to sleep. Instead, I let the thump of the hail and rain fill my mind until it is all I am aware of. A steady, hypnotic clatter that carries me along like a leaf on an autumn river.

I don't even notice when it stops.

11

THE MOON OF SMOKE

Chance people on the bridges peeping over the parapets into a nether sky of fog, with fog all around them, as if they were up in a balloon and hanging in the misty clouds.

(Charles Dickens, *Bleak House*)

AS MUCH AS I like sunshine, and as beneficial as it is to my mental wellbeing, I really enjoy the nothing days that cusp autumn and winter. The slow of this morning has gathered no momentum and instead dragged its heels into the early afternoon. It is dull and dank. Grey mist trails from fence posts and telegraph poles like spider web veils. All is quiet and subdued in an eerie still. Beaks are closed, wings folded, ears flattened and every whisker tucked as the world gently stalls. And waits.

I pulled on an extra pair of socks this morning. The fog may have checked it slightly but still the temperature dipped below freezing. I had to break the ice on the bird bowl and chip it away from the window edges of the car. Even now, a little after lunchtime, the ground remains solid and the tarmac slippery. I took a gentle drive here, fog lights on and careful round the corners. The car thermometer tells me it is 3 degrees Celsius and it certainly feels it. There might not be a breeze nipping at my nose or exposing the holes of my moth-eaten jumper, but the cold is cocooned within the dank and damp. Everything is wet, the air a mizzle of moisture. I'm not sure it could be classed as freezing fog, perhaps more of an icy ether, but one thing it all but guarantees is that I will not see another soul as I walk.

I will not be alone in wondering how I might have fared in another time or culture. How my personality might be received, what skills I might bring and how I would cope within a society I didn't understand. It is a purely self-indulgent muse, of course, conjured by empathy and ego, but a harmless curiosity, nevertheless. Yet almost all of the representations in which I might imagine myself are false depictions dressed through Western interpretation. Through the media of film and literature or straightforward racial or sexual bias. And weighted heavily, therefore, to be welcoming of the influence of a wise and learned white male. And on removal of those filters, when considering the actual reality, I realise I would be rather hopeless and such a situation would be utterly terrifying.

The romantic image of the Celts that I paint in my mind is based upon my own interests and likes. Their relationship with the landscape and the moon. It is an understandable stance, particularly considering that my anthropological affection was only born after I scratched away at the top few layers of natural history. And that was then fed through a rejection of popular culture and other interpretations of a 2,000-year-old calendar, allowing me to neatly skirt the brutality and barbarism and instead focus on totemism and spiritualism. Perhaps such focus might have stood me in good stead as a druid, but again I am being influenced by false representation. My idea of a clean-cut potion brewer, respected for his wisdom, being somewhat out of keeping with the blood-rites and animal-guts reality. I may have learned plenty as a child from those *Asterix* books, but a realistic portrayal of Gaulish society was not one of them. Neither, in fairness, was the depiction of the Romans, and, rather like the Monty Python line 'what did the Romans ever do for us?', my own paradox is that I likely would have survived more easily within the structure and regimen of Roman culture than the slight rawness of the Celts. Mind you, a mind like mine that tends so often to overthink would probably struggle wherever it was placed. It has certainly laboured within the society in which it does exist.

Perhaps, though, my developing interest in the Celts comes not just because of the lack of knowledge of their culture, but because what is known is so alien. The influence of the Romans (and Greeks and other ancient civilisations) upon modern society is marked. From sanitation to irrigation and public order, certain familiar facets of modern life were formed, in part, through autocracy.

I am not going to cast scorn upon any of the things that have benefitted the world in which I live, but where there is familiarity of structure there seems less opportunity for escapism. I have always been excited by fantastical tales, of immersing myself within the world of Tolkien or Arthurian legend. Treading where the ground holds obvious secrets is inspiring. It is the not knowing that allows my imagination the freedom to engage and create a sense of what might be. The exact same draw pulls me to water and the wilder places, mountain, heathland and scrub. There is an allure within the unfamiliar but also a respect for the things that can survive where I cannot. And walking on Eggardon at this time of year, with a hat on and a coat pulled tight, brings a deepening level of respect for those who once lived here. Intensified by the lie of the land – the ramparts and ditches and sheer ruggedness. One reason, perhaps, why I find myself drawn to certain places only when their appearance – their essence – has been tempered by the weather.

These meadows where I am about to walk are far more popular in the spring and summer, when the traditional field systems are thick with wildflowers and pollinators. The hedgerows, coppices and thickets full of birdsong. It is an amazing habitat – 180 hectares managed by the Dorset Wildlife Trust – but while it reflects a time now gone, it is also a place that feels familiar. I do like to wait for the solitude of late autumn, but more importantly I like the feel of these fields now. Everything is tired and stretched with some less obvious life-forms taking centre stage. And this weather accentuates that sense; it makes the place feel less managed and less contrived. It also throws up a rather interesting quandary. The list of chapter headings that I settled upon before writing this book has remained intact until now. This was originally going to be 'The Dark Moon', but now that it has arrived I cannot help but change it.

Fire has long been an integral part of human society, as a source of heat, cooking, protection, communication, weaponry, agriculture, energy and social interaction. For the Celts, as I mentioned in the last chapter, it formed a pivotal part of ritual and rites, a pattern familiar across many cultures and religions. From Hinduism and Christianity to Aboriginal and Native American custom, fire is revered and integral, something to be lauded and worshipped. Perhaps then, while giving such credence to Xavier Delamarre's translations of the Celtic months, I shouldn't have avoided the entry for *Dumannios* as though it were a drunken bore on a bar stool. Delamarre pointed to the Sanskrit *dhumah* and the Lithuanian word *dumai* – both meaning 'smoke'. Lithuanian is an important source for linguists, having retained aspects of grammar and phonology that link to Sanskrit and other ancient languages, and its weight in this instance is considerable. But I had fallen into a trap of accepted order and if I found agreement in one particular place then I would follow it to the bottom. And 'The Dark Moon' felt appropriate, with days brief and the sun so low when it does reach skyward. Except the moon itself is far from dark – in fact, it is surely at its brightest and most conspicuous, hence the widespread awareness of it at this time of year.

So it isn't until I have reached *Dumannios* that I find myself questioning its meaning – and that is no bad thing. I may not be solving any riddles or providing any kind of definitive reference, but I am asking questions and doubting what is accepted. That suggests a modicum of understanding just so long as I can justify my thoughts. I'm back to the Prufrock paradox again.

The month of 'fumigation' seems to be the direction that Delamarre (and others) have followed for *Dumannios* and there is plenty of sense in that. The embers of the ritualistic fires that burned through Samhain will still be warm, and through the Dark Days, fire would have been as vital to life as food and drink. But is flame not more important than smoke? A clean heat is more important than a smoky smoulder. Unless 'fumigation' should be regarded as something symbolic, rather like the singing and chanting of *Cantlos*. Perhaps I'm still filling in gaps where there are none to fill, or perhaps the answer is staring me in the face. Literally.

I walk quite quickly through the first meadow. It is only narrow, as many here are, and it isn't without character, but I always like to put a bit of space between me and the car park. This place is called Kingcombe Meadows, and at the centre is a centre – a collection of buildings and market garden that offers workshops, cream teas, and bed and breakfast. As a result, there are always people on site, whereas the meadows themselves, now designated as a National Nature Reserve, tend to be less busy. A footbridge takes me over a ditch and across an invisible threshold of my own design, and immediately my breathing slows. It is odd, really, but I feel more able to dawdle and prod and poke when I have my own space, and this little meadow is a fascination. Bound in by brambled hedges and dissected by a soggy but narrow dip, the first half is a mass of anthills, neatly spaced and each one topped with thick-stemmed grasses that splay out like a well-used paintbrush – or perhaps the hair of an errant prime minister. Beyond the dip the ground flattens as the small lea tapers and the brambles encroach, but I pause among the anthills, treading carefully between them.

They are the work of yellow meadow ants, none of which are visible as they are a species that spends most of its life below the ground. Having flown, a new queen will burrow into the soil and lay eggs which then hatch into worker ants. They begin to excavate a nest which, over time, will build up above the surface and offer room to regulate heat and moisture inside. The eggs, for example, will be moved around the nest in order to optimise warmth and cool at various points in the year. It takes between ten and twenty years for a hill to form fully, with the structure aided by the roots and rhizomes (underground 'stalks') of plants. With so many tunnels, the hill itself has the feel of a giant sponge and the wobble of a well-set jelly. Touching it, though, can collapse the myriad tunnels within, so I am careful as I step and certainly wouldn't use one as a seat – no matter how comfortable it might appear. It is assumed that new queens will displace or replace the old, as a single hill might remain occupied for 150 years or more, and each one is a microcosm, with different grasses and herbaceous plants knitting themselves into the rich, well-oxygenated mass. The worker ants encourage root spread by defecating into specific cracks or holes into which the roots are then drawn, the plants in turn creating a home for aphids and other invertebrates upon which the ants feed. In some areas of limestone downland, the anthills may play home to the caterpillars and chrysalises of the chalkhill blue butterfly, who share a relationship similar to that of the large blue butterfly's relationship with red ants. The yellow meadows feed upon the amino-rich honeydew that the caterpillars secrete, while offering protection in return. There might not be any chalkhill blues at this site, but it is nevertheless extraordinary to consider the symbiotic complexities occurring inside each mound of earth. And this colony is fairly small.

I took a lovely walk with writer and broadcaster John Wright recently, as we recorded a podcast for *Countryfile Magazine*. John took me to a site very dear to him and unknown to me where a south-east-facing hillside was absolutely covered with anthills. John had once counted them and found there to be more than 1,000, and the influence upon the natural history of that place was even more greatly marked. The hills are thick with helianthemum – the rock rose – which as a 'woody' plant is able to play host to mycorrhizal species of fungi. Mushrooms that might normally be found only in woodland appear each autumn on open grassland.

Such habitat is a stark reminder of how depleted so much of our landscape is. The government's *State of Nature Report* in 2013 suggested that 97 per cent of wildflower meadows in the UK have been lost since 1930 and one in five flowering plant species is threatened with extinction. There is no easy fix, either. The meadows here at Kingcombe represent a typical Victorian farmland landscape: small fields, mature hedgerows and no fertilisers or pesticides impacting the pollinators and natural plant growth. The result is a bucolic idyll, but not a scene that has any financial viability amid modern demand. The pasture here is grazed by sheep and cattle but relatively lightly, allowing the habitat to be maintained rather than any reliance being placed upon the stock for a return. It would be impossible to farm in such a manner on a wider scale, as sad as that may sound.

What I can do, though, is appreciate this place for what it is, although my sense of connection is not particularly profound.

I wonder if, this being a snapshot of a relatively short period of time, there isn't the soul beneath my feet. The conservation of this habitat is a wonderful thing, but I feel little for the people that would have originally created it. I can visualise them working, but that is through the eyes of others. Literature from the likes of Thomas Hardy or Richard Jefferies, artists like John Constable or Claude Monet. Not necessarily works that I admire, but imagery that we are all exposed to, often through film and television adaptation. As much as this is absorbed subconsciously, so I probably push back against it without consciously doing so. A rejection in part of popular culture or theme that can be difficult to negate, much as I put up mental barriers against the bombardment of Christmas schmaltz and advertising but forget to lower them at Christmas itself.

It has to be said, though, that there is a somewhat Dickensian edge to today's rural scene. As I have stepped into the next field, the trees around the edge smudge in and out of view as the fog breathes around them. There is a bleak beauty to this ether, and such subdued sound. The squelch and crunch of my footsteps as they crackle through the ice and mud seem amplified when they ought to be muffled. The great grey blanket bouncing the slightest sound back rather than quashing the waves. This isn't the smog of Victorian London – the air is clean for one thing – but there is a touch of the Gothic to the day. I really like it.

There are two oak trees that mark the top boundary of this meadow. They stand on a small ridge that looks to have been formed from excavating the dip that runs beneath it. Today, it is a fairly unremarkable fold that would be easy to step across without giving thought, but given its direction across the fall of the land, it was once something deliberately more substantial. My first thought is a ha-ha or *saut-de-loup* – a ditch built in place of a fence or with a barrier placed within it. A design originating in France, as the names suggest, with the 'ha-ha' allegedly stemming from the reaction of a son of Louis XIV upon being warned of the impending drop. A more feasible, but less fun, theory suggests the term was an adaption of 'half and half', it being part fence and part ditch. They became a familiar feature in deer parks and estate gardens as they allowed the enclosure or obstruction of animals without blotting the view. A sense of continuation in aspect from the grand gardens into the deer-dotted pasture beyond.

I am not aware of this area ever being a deer park, though, and nor is there a stately house from where people might have felt their view spoiled by a fence line. More likely, then, is that this was a drainage ditch leading run-off from the rise beyond away from the field I have just crossed. The presence of club-rush across the lie behind me is evidence of the moisture within the soil, and there is a fair squelch should you step among the deep green of the stems. When dug, the ditch would have been more substantial and in turn effective, and, given the position and age of the oaks (that wouldn't have been dug around), this dates back at least 150 years.

'Crack!'

It sounded for a moment as though a branch was breaking, but in the deathly still the detach and fall of a single browned leaf clanks and crackles earthward like a spluttering two-stroke engine. I make a move to catch it, but even a dive wouldn't get me close. The instinct is there, though, the infuriating fun of grabbing falling leaves. I must catch twelve each autumn in order to secure good fortune through each month of the forthcoming year. A tradition I didn't begin until my late 20s, having spent far too long in the company of the Yates family. Now it has become a mild obsession, but I have already bagged my dozen for this season, despite a slow start. And the oak is a reliant tree from which to complete the set. It hangs on to its leaves later than most and, as just happened, will drop them due to cold and without the blow of the wind.

I wait for a few moments, just in case another leaf falls, and it proves serendipitous. From my right comes a familiar buzz, and the russet and yellow of a hornet swings lazily past my nose. It is a cold day to be an insect on the wing, but it is not unusual to see hornets this late. She is a queen, an inch and a half long and likely looking for a place to hibernate, swinging around the limbs of the oak beside me, a slow swagger of someone ready for their bed. They are common locally, which is a treat because they are a beautiful, and invariably docile, animal. Happy to manoeuvre around you, even beside their nest, and content to sit on a finger or palm should the opportunity present itself. The sound is most evocative, though. A low resonant hum that floats dreamily through the quiet of a late autumn woodland. It reminds me of the timbre of a cello cutting through a sea of violins as a piece drops into a reflective movement. A change of mood and tightening in the stomach. Sleep well this winter, my queen.

The fog has thickened slightly, leaving a faint mist across the lenses of my glasses. And as I pause again to look around me, every hedgerow and tree is obscured. I know this field well, though, and change course slightly to check on a favourite mushroom ring. A circle of parasols some 30 yards in diameter that have the presence similar to standing stones. Not today, though; the remnants remain, but the frosts and ice have left a round of blackened, soggy dollops. I veer back to my original path, keen to see what else might be lying around my feet, but even the inkcaps are going over. There are one or two tiny little toadstools, nameless to me although I'm sure John Wright would rattle off the Latin names. I will find waxcaps a couple of meadows away, but the fungi in this field have served their season.

Through another gate and then right through a proper welly-grabbing squelch. I almost lose a boot and get plenty of splatter up my thighs before I duck through a trail of hazel branches, stubbed with the green of freshly formed catkins, and reach the browned bracken sanctuary beyond. A nuthatch bubbles from the thicker set of trees to my right, but all else remains quiet. Other birds are moving in the branches – when I stop, the sound of clawed feet gently scratches from all around me – but energy is precious and voices hushed.

As I near the top of the bracken-brown meadow, I see movement behind the hedge. Nothing untoward or unexpected, but sheep are grazing and it might not be kind to walk among them in the fog. I like to think I maintain a relatively benign presence among livestock, and not having a dog certainly aids that, but in this gently swirling blur the sheep might be easily spooked. Instead, I veer right to a gate I do not normally pass through which will cut my route short but only by a single meadow. As I edge alongside the hedgerow, I catch the lightly sweetened smell of sheep-shit with a lanolin edge. It's rather glorious, whisking me away to the uplands of Wales or Cumbria. And while I won't dwell now on the conflict around sheep farming, excessive grazing and rewilding, I will let my inner child enjoy the escape.

A thought comes too, about the month and the moon and what actually is staring me in the face. The Moon of Smoke – it needn't be literal and, although the importance of fire is undoubted, this is the period of the year when fog and mist are most prevalent. And without knowing how it is formed, what assumptions might be made? Perhaps that the ground itself is smouldering as the life upon it withers and dies. A slow burn that browns leaves and curls bracken. A symbolic cleansing.

Last winter I went fishing on the Dorset Frome, a lovely chalk stream that sparkles through the Dorset landscape before meeting the sea at Poole Harbour. The short day rattled past and my anticipation was building for the final throes, the witching hour, when fish are often stirred the most. Low light levels provide opportunity for surprise and a stretch of river that ran lifeless for hours can come alive. It is my favourite part of an angling day. But having caught a nice grayling just after sunset and watched it slip back into the current, I had a sudden urge to pack up my rod and simply sit. I was on a wide floodplain on the lower river, alone and with no one else in sight. Birds were moving to roost and a nearby run of alders were silhouetted like shadow puppets against the orange glow of the south-western sky. And as I sat and watched, a mist simply appeared around me. Just a few feet high but in one moment there was nothing while in the next there was a candyfloss veil. If it wasn't magic then it was something beyond comprehension, and had I not known of the rapidly cooling water droplets in the air I would have presumed it something utterly momentous.

And now, as I walk and breathe through this odourless smoke, I cannot contemplate any other meaning for *Dumannios*. My thoughts are no doubt swayed by the moment, but direct association can also make a clearer sense of something. And I don't doubt that my theory will have been made elsewhere and given all of the writings I have dipped into there is a chance that I have been unwittingly prompted by another, but that doesn't really matter. I make a nod of apology for overlooking Xavier Delamarre's deduction and walk on. The Moon of Smoke seems perfectly apt.

Each one of these meadows has a character of its own, but this field has many to itself. It is one of the larger meadows on the reserve but that does not detract from its intimacy. It is shaped in a sloping dog-leg, folding around a large coppice. I will walk a vague loop, heading along the eastern hedgerow to the point where I would have entered (but for the sheep) before dropping halfway down the slope and coming back. The further corner and the meadows beyond hold little allure. I wonder if they are an area more recently acquired and, therefore, less rich in depth and life. The feeling is similar to that of the stubbled field behind the hedge on my walk beneath the Harvest Moon, so perhaps it is a staleness that I sense.

In hugging the hedgerow, I spook a fieldfare who flies with a brash 'chak-chak-chak!', in turn panicking several dozen who dot the higher branches, obscured to me by the fog. They pour away, striking calls with sharp wing beats to match. They are not so woodpecker-like in flight as a mistle thrush but have a slight undulation as they flap and fold. I offer an apology. I would not have walked so close had I seen them, but it also seems that their movement and voice have triggered something of a chain reaction. A blackbird shrieks its scarpered alarm, and a magpie replies with a rattle. And then a stranger 'plip, plip' – like a small pebble dropped into a shallow well but with an added metallic ring. If I hadn't heard it before I would struggle to guess the source and even now, being unable to see the bird who has made it, I am grateful for the 'swish, swish' of wings in the fug overhead that confirms it. Somewhere in the gloom a raven is on the move.

'Braak!' There is a second bird, and this call is more guttural and typical. Ravens have an extensive vocabulary and occasionally I can persuade one into conversation. The first effort needs to be good, close enough to the raven to pique their interest. A bungled, blatantly human impersonation – such as I just garbled – will be roundly and rightly ignored. They are curious birds and although wary, will also come close to investigate if something seems out of place. More than once, while crouching in an open field poking a mushroom or flower, I have had a raven drop within feet of me. The moment I move they sweep away, but they are apparently curious as to my motive and perhaps even checking whether I am a potential meal. Carrion is a principal source of food for the raven, although they are unfussy and will eat seeds and grains or hunt small animals. Their intelligence is their primary weapon, though, and possibly why they are viewed by humankind with such suspicion. Their reputation for attacking livestock is perhaps exaggerated, and what little scientific research has been made in the area (more specifically with crows) suggests that they target animals that are already dead or ailing. This is, of course, little consolation to the shepherd who has to deliver or euthanise a partially eaten lamb even if it may have been moribund before the raven's attack.

The problem seems to be that when we look at a raven we see the thought behind its eyes. It is a problem solver, a strategist, a bird that doesn't automatically take flight when approached but will consider the situation first. That is a quality that some people struggle with and they in turn apply anthropomorphic intention and behaviour that is not necessarily warranted. Rather like the fox in a chicken coop that is falsely convicted of blood lust, the raven is an opportunist that is hated for its habits, and more so because we apply unpleasant human traits to them. Then we tend to see what we think we are seeing rather than what is actually happening. A fox will kill more than it can eat because it will cache the food and not have to hunt for several weeks. The raven will loiter at lambing time because there may be protein-rich pickings from the discard. Attacking a live animal comes with the risk of injury and even minor damage to a wing could cost a raven its life. Risks will be taken when food is scarce, though, and if that action proves successful then the behaviour might be repeated and copied.

Another factor that is understandably concerning livestock owners is the increase in population of the raven. It is a species that, while having been historically widespread, was eliminated from most lowland areas, predominantly through persecution, in the nineteenth century and restricted to the uplands and extremes. Today, with protections in place, it has recolonised much like the buzzard has. Ravens now occur across the British Isles and, while they are far from common, there could come a point where their number becomes problematic. And, unlike the buzzard, their adaptability of diet and greater intelligence might allow them greater expansion than would naturally check the population of a top predator.

For now, though, I can enjoy my interactions with them without dwelling upon potential conflict. Their intelligence merely adds to their appeal, and we have been fortunate at home to be able to watch them throughout the year, although it is the upcoming courtship that is most fascinating. The local pair use the east ridge for lift and will spend hours cementing their bond. They fly so closely together at times that it is difficult to separate them visually and the male (I presume it is the male pursuing but there is little to distinguish the sexes) will move behind the female and lightly pull her tail with his beak. So far this autumn they have been playing a game on one of the fence posts, where one perches on the post itself with the other on the ground. They look at one another and move their heads before taking brief flight and swapping positions.

Many bird species have distinct and sometimes spectacular courtship rituals, but few are as individualistic as ravens, who seem to develop idiosyncrasies unique to each pair. Synchronised flying is a common behaviour, known as 'unison flight', as are the 'flight-rolls' that see a bird tucking in its wings and flipping upside down, but the tail tugging and post playing far less so. 'Our' pair of ravens have apparently cultivated a deeper intimacy over time. Not just reacquainting for the bond of breeding but sharing their own little moments of fondness.

The huge black bill of a raven is an unlikely looking tool of tenderness, though. They are big, powerful birds and formidable in appearance. It is unsurprising, therefore, that they are a species incorporated within the mythology of many cultures. In Norse legend, the god Odin was accompanied by two ravens, Huginn and Muninn, who were gifted with the ability to speak. They would fly out at dawn every day and travel throughout the world gathering information which they would then share with Odin upon their dusk return. Depictions of Odin often include the two ravens, alongside two wolves, Geri and Freki. The presence of both ravens and wolves is interesting, because the two species have a long association. Ravens are often the first species to visit a wolf kill, having apparently followed the hunt. Some witnesses have suggested the wolves themselves will respond to the call of the ravens, as though the raven is the wolf's eye in the sky alerting the pack to prey. Another theory suggests that the principal reason wolves hunt as a pack is to ensure they can protect a kill from the ravenous ravens, but this feels like a thought too far. Aside from the hierarchical and behavioural processes of pack formation and mentality through most canine species is the fact that ravens, though powerful, will not break down large prey such as deer or elk before a wolf's teeth have torn a beginning.

They are undeniably swift to identify a food source, though, and have long been associated with scenes of battle. Drawn by the bloody pickings, the ravens were seen as portents of death or perhaps the departed souls of the slain. Irish legend points to the goddess Morrígan, who took the form of a raven as she soared across the battlefield, while in Wales it was the giant Brân, whose name translates as 'raven or crow', who, in a battle against the Irish, sacrificed himself to ensure victory. As he lay dying, he implored his men to cut off his head and take it to London, there to bury it on the White Hill where he would ward off any future invasion. White Hill is believed to be the location of the Tower of London, where ravens are kept, with similar promise, to this day.

The ravens in the fog have moved off to the south, where I know from memory there is higher ground and trees where they might perhaps hold territory. The odd 'krakk' and 'kronk' still echo through the grey, although it sounds as though they are the notes of casual, seated conversation rather than anything to do with my presence. I have dropped down to a lower slope of this meadow having reached the point where I normally enter. A cattle trough marks almost perfectly the apex of the dog-leg and from there I walk down to the bracken line, a dead and brown swathe of which wraps the centre slope, and begin to work the homeward path. A lone oak stands in the basin of this curve and as I approach it slips in and out of the grey as though playing arboreal peek-a-boo. Its branches are less clothed than the two oaks that drew the hornet's attention, most leaves already fallen. It is a similar age and size, but stands wizened and skeletal, hopefully not a sign of disease or poor health. My interest, though, is soon snatched by a soft glow among the grass swards beside my feet. One reason I always visit this meadow and take this route is for the fungi. Kingcombe is renowned for its waxcaps, playing home to more than two dozen species, and this section of slope seems to show them off as well as anywhere. The one beside my foot is a crimson waxcap (I'm fairly certain – but where is John Wright when you need him?) and is an impressive size. It is not as wide as the span of my hand but considerably larger than the palm and positively enormous compared to the snowy waxcaps I sought beneath the Quiet Moon. And thanks to the elevation of the ground, I can position myself below it and get a good look underneath the cap without risking damage. The stipe and gills are more orange than the rich red of the cap but are equally impressive. And the glow comes from the waxy, candied coating that doesn't look altogether real. On a sunny day I like to poke my nose tight beneath them and look up through the cap, the diffusion of light almost tangible enough to taste.

As I stand back and look, so the waxcaps emerge from every tussock and fold. They might not glisten as they would in the sunlight, but through the grey veil the colour seems almost artificial. Standing proud like the symbolic red flag in Sergei Eisenstein's 1925 film *Battleship Potemkin*. It is as though someone has taken a paintbrush to the hillside as Eisenstein took to the celluloid. And there is a glut of other *Hygrocybe* species too. Slightly smaller scarlet waxcaps, big clutches of snowy and meadow waxcaps, two parrot waxcaps in glorious green and others I would only be guessing the name of. The meadow has delivered as it always does and it looks as eerily beautiful beneath the Moon of Smoke as it might in the sharp of a crisp, cloudless early winter day.

As a postscript to the prevalence of fog in late autumn, I contacted the Met Office to be reassured as to my thinking. They kindly responded but told me that, although difficult to measure, January is the 'foggiest month', albeit by fine margins. This obviously wasn't the reply I had hoped for, but before I scratched my theory behind the Moon of Smoke, I considered the change in climate. Two thousand years ago, the weather in January would have been considerably colder than today. People stayed at home through *Anagantios* in order to avoid the great freeze rather than the swirl of fog, so perhaps *Dumannios* was indeed the time of smoke. I'm sticking with it in any case.

12

THE COLD MOON

Cold in the earth – and fifteen wild Decembers,
From those brown hills, have melted into spring.

(Emily Brontë, 'Remembrance')

BELOW ME, to the north, I hear the march. The rhythmic clunk of metal and sharp thud of hardened leather on well-worn flagstones. The echo is slight. Deadened by the thick air that bleeds the edges of the Cold Moon in the east. The breeze brings a chill from the north, but it is light. Almost breathless in the most part, and for several moments, as my ears adjust to the still, the legion of foot soldiers seems to be approaching at speed.

'Beware the Romans ...' came the warning, should I walk on Eggardon after dark.

Perhaps I should have heeded those words rather than come here in the murk of a winter night. And although I had meant to do this, to finally walk by the light of a full moon, I cannot help but feel on edge. I ought to be able to rationalise this, and as that thought comes to mind so I do in fact feel reassured. Of course, I don't believe in such things. Not really, though I try to remain open-minded whenever I am able. There are countless more things in this world that I do not understand than those that I can wholly explain. Perhaps this is actually the moment I discover something new – to me, at least. Perhaps the warning carried some weight. Because now I am here, alone, and with a ghostly Roman army marching towards me through the purple ink, my timing seems to have been impeccable.

I breathe out slowly and close my eyes, pausing for a moment before inhaling deeply and deliberately. I wonder quite where the soldiers would be walking. The northern sweep is far too steep to maintain a march, and besides, even the Romans wouldn't have built a road on such a gradient. Of course, ghosts, being ghosts, might be moving through the air itself, treading a path that itself was lost a millennium and a half since. Or perhaps I have been completely disorientated by the darkness and a fervent imagination. Perhaps I'm actually hearing an echo of the true source.

Or perhaps I am listening to the sound of a slightly worn pump on a cattle trough.

I smile. I walked here a few days ago with Sue and we paused at almost this exact point to watch a kestrel. It rested briefly in a wind-stunted ash tree up ahead of us, before lifting as perfectly as Hopkins's 'Windhover', hanging on the breeze no more than 50 yards from us. Our eyes on equal level, but the falcon's gaze fixed far beneath our feet.

I had lifted my binoculars for a closer look at that stilled stare. The head utterly unmoving as the feathers around it danced to the wind, riding the rein of Hopkins's 'wimpling wing'. And as it fell from one watchpoint and I followed its path as it dropped to swing along the ridge to find another, the slate grey of a cattle trough flashed across the lens far below. Noticeable only because it cut such a contrast with the faded tones of winter, but peculiar nevertheless that my mind should have logged the image. After all, I didn't know then that I might need rational reassurance now.

It isn't the first time I have been caught out by the darkness. By mischief made in my own mind. But with the source of the sound suitably isolated, it shrinks back into the background. It sits there with the low growl of a tractor that rolls up from the fields beyond Eggardon's bow, although the bulk of the hill does block out any sound of the A35. Not that there will be much traffic rattling along it this evening. It isn't late, but it is a Sunday and we are less than a week away from Christmas Day. Most people will be where they need to be on the eve of a Monday and more will have cancelled plans that they may have had. With a new and more infectious Covid variant on the march, the threat of virus-enforced isolation or another wholescale Christmas lockdown has prompted widespread self-imposed house arrest. Bad news for the hospitality industry seeing cancellations at the peak of their working year to follow almost two years of hardship, but understandable given the circumstances. Sue and I certainly don't wish to jeopardise our plans. It is two years since we last made the short hop down to Torbay, to stay with Sue's mum and see her sister and her three children. And while we enjoyed last Christmas for what it was, the two of us tucked away with fires and films, we are so excited to be heading west for this one. Lots of games, lots of banter and lots of laughter. And a bottle of single malt waiting for me beneath the tree from my mother-in-law – what could be better?

It's also a time of year when we are constantly reminded that we do not have children of our own. The endless bombardment of sparkle and shine is heavily weighted towards kids, and while we do our best not to dwell, it is also nice to know we can enjoy the company of someone else's. So it is that towards the end of the week, alongside millions of others, we will be poking swabs up our noses and hoping for just a single red line. And we should be fine. One advantage of not having children is being able to avoid the germs that rattle around classrooms and playgrounds. Life in lockdown is for us not so different from life in general, but the impact that isolation has had on some is severe. Throughout the last twenty-two months we have embraced the landscape around us with a deeper sense of appreciation. Our circumstances no longer a hindrance or unhappiness but an unexpected position of wealth with the carpet having been pulled out from beneath all of our feet. Money and material are no use when all we crave is fresh air and open space.

I have laboured this point before, but another Covid surge brings a timely reminder of it. And as other people are locking their doors and hoping, I have been able to pop up the hill in the dark and watch the full moon glisten above an ancient hill fort. Not that there is much of a glisten at this moment. Broken cloud drifts slowly on that light breeze, masking the moon, although the edges glow from the force of light behind and my eyes are adjusting well to the darkness. Not that I have actually ventured up on to the fort itself. I am walking the tiny lane that runs along a ridge on the northern edge of the great anvil; it makes for safer footfall and my presence is less likely to panic any livestock unused to seeing people after dark. There have been several incidents in recent months where dogs have been unleashed and subsequently chased the cattle and sheep. A heavily pregnant Highland cow was chased off the edge and fell to her death, but despite the publicity and signage still people let loose their dogs to run amok. They are a friendly herd and love a scratch, but it seems irresponsible to step among them at night, especially when their night vision will be so much better than mine. I have stumbled across fields before as I squint to find a path back to the car after a day on the riverbank, and my ineptitude seems to stir bovine interest as much as the fishing rods on my back and the bait bucket in my hand. A slip or trip could easily spark alarm and in an instant a situation of mild-natured curiosity becomes a threat to life.

There is another reason I am walking on the road. The devil in my mind that conjured a marching Roman legion had, for many years, an overwhelming grip once the sun had set. Darkness would terrify me. Deep into adulthood I had dreaded the dark just as those who are phobic about spiders and snakes. It was irrational, embarrassing and debilitating, and not aided by those couple of instances when danger did come at me from the black and present actual physical harm. It was all so real, and not something I could simply snap out of. To understand it I applied a similar approach to the methods that helped me give up smoking. Digging into the roots and questioning my mind's own conviction. To beat the nicotine addiction, I had to tease out every schema that had lodged as a false memory and relearn it. Similarly, in poking around my own head I realised that my fear was less about physical threat and far more about isolation. The insomnia that accompanies depression and anxiety will compound both by leaving a mind lost in darkness. Daylight would come as a respite rather than a resolution, grounding in the form of birdsong or float of clouds. Distraction by television or radio. Until the sun dipped and the fear grew once more.

I don't wholly embrace the dark now, but my relationship with it has improved immeasurably. There will always come moments, though, just as an ex-smoker can never fully let down their guard. A waft of cigarette smoke in the sunshine can spark a brief urge, but it subsides just as quickly. Darkness is more consuming, though, and a stumble more likely to become a fall. But letting yourself go just a little bit can turn a potential trip into a moment of excitement. There was a stable part of my mind that instantly rationalised the sound of the water pump that I heard a short time since. It didn't immediately identify it, but it was assured in its interpretation. It almost allowed my playful mind the freedom to run away with itself while it determined the source. And so it did – conjuring a ghostly legion and almost willing it to be true because there was a deep assurance that it wasn't.

I look over at the main bulk of the fort, the rim of the highest ramparts surprisingly sharp against the subdued light of the cloud-masked moon. Were I to slip back in time a couple of thousand years, I wonder what activity there might be up there – within the fortifications? If people were present then, would they mingle after dark? Would they keep warm beside communal fires or rely on the snug of their beds? Firewood, although plentiful, would have been a vital resource – keeping it dry, critical. It would have been used sparingly and as though lives depended upon it, because that was the reality. Not that lighting a fire within a confined space wasn't without its risks.

In 2010 a metal detectorist discovered a copper alloy mirror in a field close to the village of Langton Herring around 10 miles south-east of Eggardon. Though remarkable in itself, with a grip formed from a single loop with tear-shaped void and unique pattern, the mirror was placed on the chest of a young woman, believed to be aged between 19 and 24 and, given the style of burial, likely to be of noble blood. Archaeologists estimate that she was interred between AD 25 and AD 53 alongside artefacts from diverse sources, reflecting her importance and also the shift in culture that was evident prior to the Roman occupation. That societal advancement may also have contributed to her poor health, with 'areas of fibrous woven bone formation' suggesting that she suffered from an upper respiratory tract infection consistent with excess inhalation of smoke. It seems likely that as house-building became more efficient and properties more airtight, people's health suffered as a result, the fires lit for cooking and heating also delivering a dangerous amount of smoke. So the Durotriges on Eggardon may well have been tucked up, but not necessarily to their long-term benefit.

The weather would likely not have been as benign as I am feeling this evening. It is chilly, but not cold. There might be a brushing of frost by morning but water won't freeze and there is no lying snow or risk of any falling. Nudge the temperature down a couple of degrees, though, and the world would look and feel very different. The low-pressure system that is offering the possibility of a white Christmas for those further north would dump snow, rather than rain, here too. It is hard to imagine that the Durotriges would be anywhere other than their beds right now. Tucked up conserving energy if not actually asleep. A lifestyle built around natural rhythm and more reflective of that of other mammalian species with which they shared the landscape. They might not have hibernated as such, though recent theories suggest that ancient humans may have. Examination of fossilised remains uncovered at Sima de los Huesos (the Pit of Bones) at Atapuerca in northern Spain has found similarities with that of cave bears and other hibernating animals. Lesions on the bones point to a disruption of bone growth for several months of each year consistent with hibernation, but, if true, this would have been primitive humankind some 400,000 years ago. A being in the thick of evolutionary change. Other cultures, such as the Inuit, who survive the cold and dark of Arctic winters, do so through an adapted diet: a variant gene that helps their bodies to store the high fat content from a diet dominated by marine animals – a similar specialism that occurs in polar bears. But simple behavioural traits among the Inuit and across all societies can impact metabolism. Many of us will be weighing more in a fortnight as the new solar year begins. A result of overindulgence through the festive period but also of inactivity. In the far north, where the days shrink into winter before ceasing to be days at all, human activity will lessen, particularly with the dangers of darkness. Polar bears will actively hunt humans as a food source, so the threat of darkness is altogether real. As it might have been for the Celts on Eggardon.

Brown bears are believed to have become extinct in the British Isles around 1,000 years ago, while wolves clung on until the mid-eighteenth century. Both species would have been wary of humankind, but would also have posed a serious threat, particularly if starving, startled or protecting young. And in the dark, humans would have been at a distinct disadvantage. The make-up of photoreceptor cells in our eyes might give us better perception of colour but require a high level of light in order to work. A good reason to avoid the use of a torch or other artificial light source after dark, and also, perhaps, a reason why fires wouldn't have been kept burning on the hill fort overnight. Aside from drawing unwanted attention, there would have been the impact that the light of the flames had upon those tasked with looking out for danger.

My eyes have adjusted fairly well as I have walked, but my night vision would be far sharper had they been exposed to the gradual loss of daylight. When fishing, I can watch my float-tip through dusk as it smudges into nothing. When I finally look away, I am surprised by the darkness that surrounds me, but that slow adjustment gives me sight at a time when I would normally flail. Tonight, walking on tarmac certainly helps: knowing that the ground beneath my feet is firm enables me to apply focus elsewhere. Not that this lane is without hazard. It is narrow and peaked, a ditch on the northern side and a 150-foot tumble down the other. The barbed-wire fences might break a person's fall but wouldn't halt a vehicle, and drivers are sensibly tentative as a result. It is also one of those glorious stretches of country lane that is broken and chipped, the middle mossed and grassed. In the summer, the bloom of grassland flowers that shimmer on the hill itself has a smaller, but more defiant, show along the lane. Windblown seeds finding foothold in the wear. It is a route most dramatic if approached from the west. A long drag through high hedges and then a slow meander around Eggardon's toes, before a sharp, low-gear climb up on to the top. There, the world to the north opens up beneath you, pasture and trees that flow into the thicker scrub and woodland of Powerstock Common – where I sought the song of willow warblers beneath the Growing Moon in early spring. It was also the direction I came one evening a few years ago, a little earlier than I am here today.

My lights were on full beam, but I would have seen the owl without them. It lifted from the fenced depression to my right, drifting 50 yards or so along the edge of the lane before settling on a post. I stopped the car and watched, the headlights softening the rufous into yellow and the light browns into white, the eyes flashing as they looked straight at me.

I had waited six years to see a short-eared owl locally. The rugged landscape makes for ideal winter hunting, and I learned soon after arriving in Dorset that they were seen in the area, albeit occasionally. But there is almost too much good country, too many places to look, and the prospect of finding a short-eared owl by design was daunting. My propensity for keeping my eyes open and trusting to fortune will regularly deliver moments of serendipity, but each winter passed without the slightest whiff of a shortie.

I travelled elsewhere to see them. Regular winter roosts are found in the north and extreme south of Dorset. Places that are reliable and well watched, and when you know that you are in the right place, you find deep reserves of patience. But it seems that even in the summer, when I've visited Scotland or other areas where short-eared owls breed, they are not a bird I seem to stumble on without searching. Their tendency to tuck up in heather or long grass, often on their bellies with a round face forward like a content cat, means they are almost impossible to spot unless they are on the move. And a bird that typically nests on the ground tends to be rather good at concealing itself.

Having spotted the short-eared owl on this road, I returned at dusk on a number of occasions in the hope of a second glimpse. There was no further sign, and I was reminded of just how awkward the geography of Eggardon can be for a wildlife watcher. I could walk one side while goodness knows what might be cavorting on the other. Ten owls could tuck up among the tussocks of the western slopes and remain undisturbed until spring.

But while the shortie eluded me, I did secure views of a rather lovely consolation prize. On several occasions, just at the point where the lane drops from the fort edge and snakes its way down towards Little Dorset, a ghostly shape drifted north above my head. Much as this habitat is apt for the short-eared, so it also suits the barn owl.

I was fortunate as a child, having grown up in a valley where barn owls were a familiar sight. They've largely gone from that area now, the vast, open arable fields providing little food, while the old barns and outbuildings, where they seek shelter and nesting opportunity, have been tidied or replaced.

They are far from common in West Dorset, despite the seemingly hospitable habitat, but remain a familiar, if irregular, night-time sight – particularly in winter. They are also a bird that as an angler I benefit from seeing. The meadowlands that fill our river valleys are often filled with voles, a favourite food of owls, and the paths of anglers and barn owls frequently cross – particularly in crepuscular light. And it is in those moments, especially at dusk when those photoreceptor cells begin to struggle with the lowering light levels, that a barn owl can appear eminently spectral. This is compounded by the fact that it is our peripheral vision that tends to pick up movement in low light, and a silent, ghostly apparition might snatch your attention but disappear as soon as you look directly at it.

Just for a moment, the moon slips fully free from its veil, hanging bright and casting sharp shadows from the fence posts. My eyes are drawn straight to it, something I regret a few seconds later as it once again becomes smothered by cloud. I have to blink the exposure from my eyelids, and my night vision has been set back by fifteen minutes. If an owl drifted by now I wouldn't make it out. Instead though, almost on cue, something else whistles past low overhead. My immediate response is to reach for my phone to press record – a futile action because the bird has long gone and will not be heading back. And now, as I try to form a shape of the sound of the wingbeat, my memory is already corrupted. It seems to have been tracking close to me because there was a sudden change of tempo and clatter of wings as it changed course in response, I presume, to my presence. An image springs to mind from what my ears did grasp, though. Just a week or so since, down at Powerstock, a bird burst out from the thick of undergrowth as I passed. A mottle of browns, rufous rump and whirr of wings. I didn't need to glimpse the long beak to know I'd inadvertently flushed a woodcock. And the slightly mechanical chop of wings that just whizzed by in the black reminded me of the sound I heard then. It makes a fair bit of sense. Woodcocks tend to spend their days tucked up in cover, before moving out into open pasture or soft ground to probe for food. They are quite fond of dipping their beaks into cowpats, knowing of the invertebrates they might find within or beneath, and it could be that this bird, if it was a woodcock, had been checking the wares left by the Eggardon Highland cattle herd. Perhaps it was disturbed by the lights of the tractor and sprang a route north to wing its way back to the sanctuary of the woods.

Perhaps I am being a little assumptive because I want it to be so, but if that was a woodcock then it is quite timely. It was a species that I had meant to give mention to in the previous chapter – even if I had not seen one. The moon that I lost so completely to the fog is sometimes known as the Woodcock Moon, a moniker that has both substance and a sense of relevance to native culture. Our breeding population of woodcock is in decline. The British Trust for Ornithology surveys in 2003 and 2013 found a drop of 29 per cent to a figure of 55,000 pairs. With another national survey imminent, the results are likely to follow that downward curve.

Elsewhere, though, particularly across Russia, the breeding population is certainly more substantial even if we cannot be sure as to its stability, and those birds head west in considerable number each autumn to avoid the worst of the Siberian blast. Because of their nocturnal or crepuscular habits, this arrival is not made with fanfare or flash of feather, but a quiet and understated hush. Telltale holes in those cowpats being one of the first signs of arrival, or the zip across the sky of a stout bird as the light fades. A once popular belief was that woodcocks carried goldcrests on the arduous crossing of the North Sea. The tiny goldcrest, which, alongside the firecrest, is our smallest bird, was deemed too slight to be able to make such a journey itself, and with a similarly timed influx escaping Scandinavian or Siberian winters, logic pointed to woodcock carriage as the only possible explanation. Another long-held belief is that woodcock can carry their young in flight, and this might have a slightly greater substance. A piece in the *Irish Naturalists Journal* by J.A.S. Stendall in May 1926 described how a small party had witnessed the process. The adult was seen taking off in the shape of a ball with its head tucked beneath its breast as it kept stable the chick that it then carried between its feet. The accompanying field sketches are well worth looking out for. Other observers have asserted that the chicks are carried between the thighs, which would counter better the issue of aerodynamics. Perhaps, though, there is pertinence in the fact that more contemporary accounts, in an age of digital cameras and smartphones, are rather lacking.

Of course, the moon that keeps teasing light at me this evening has nothing to do with woodcocks or smoke, but I have again become conflicted as to its actual meaning. *Riuros*, it seemed, was one of the easier months to label. If I was accepting of its rough position in the solar year, then the translations offered seemed perfectly in order. In Old Irish, *reud* means 'great cold', in Welsh *rhew* translates as 'frost or intense cold', while in Breton *rev* is 'frost'. *Riuros* could be little other than 'Cold' – and although I am not uncomfortably so tonight, this is, and would have been, a time of coldness.

So why the doubt? I have come across alternative translations – linking to the Old Irish *remor* meaning 'big or fat' and Breton *revr* 'ass or strain' – but, if you will excuse the pitiful pun, they don't seem to carry much weight. Cold seems fairly unarguable, but it is less the meaning of *Riuros* that is now bothering me, but of *Ogronios* that is soon to come around again. The 'Moon of Ice' seems rather similar to the 'Cold Moon', and although both are apt, it seems odd that there might not be a greater distinction. It could be that there was a depth of meaning lost in the translation. That 'ice' might reflect a time of more significant solidity. Permafrost beneath frozen snowfields and cat-ice on the coast. The cold of now is still tempered by the dampness of autumn and residual heat within the ground. There might be snow and frost beneath the Cold Moon, but the serious freeze is yet to come.

And if one of the months has been mistranslated, then which is it? Perhaps it is both, or perhaps the words evolved from something similar in pronunciation but different in meaning. Through ages where words were spoken but not actually spelled out, the eventual phonetics would be somewhat subjective. Chinese whispers through Stone, Iron and Bronze. After all, we know how easily a nadder became an adder. What remains curious is that it isn't until now that I have had the confidence to question my own original thoughts, garnered as they were from the minds of others but moulded to suit my needs. It is as I have trodden deeper into the footprints left so long ago that I have begun to feel something of an understanding. It is barely a scratch at the outer layer, not nearly so qualified as the early naturalists such as Gilbert White were to assert their thoughts to bird migration or song. But unless you begin to scratch, there is nothing in your ears but the echo of others, and even the greatest of minds are prone to supposition.

I do wonder, if I were to continue to ponder the keeping of Celtic time, where my own interpretations might lead. The ice of *Ogronios* might melt within its next coming, and what if I went through the entire cycle? Five solar years of reading, research and self-reflection. It would probably take that long for me to get a decent enough grasp of the subject – certainly so that I might have confidence in my own thoughts and opinions. But that is not why I have picked up this thread. I have no wish to become expert at any one subject, and, more importantly, nor do I have the application. My mind loses interest before it gets too stretched and instead I am quite happy to scratch at the footprints that I happen to be stepping in rather than pause and dig up the singular treasure beneath.

I have paused to lean against the metal bars of a gate. This is the point where the straight of the lane plunges down the side of Eggardon's bow and I'm in something of a quandary. I haven't actually walked very far, but I have mooched at a rate far slower than I might in daylight. As a result, I am not feeling obliged to walk further for the sake of time or exercise, and the prospect of heading downhill doesn't sit well with me either. I think, rather than this unsettled sense being an aversion to the dark – and the ink does stain far deeper within the hill's shadow – it is a reflection of empathy for the fort dwellers. That's what I'm telling myself at least. Eggardon must have been a safe and secure place for the people who dwelled there. Exposed and windswept, but an island sanctuary in a sea of darkness and danger. And the sounds would have carried greater resonance, too. There was no cattle-trough pump or tractor rumble or faint hum of the main road. Instead, the sounds of the wild would have echoed: foxes and badgers, the howl of a wolf. And the belief in otherworldly spirits that enter this earth, even without the windows of Samhain wide open, could easily be stirred by the sounds of the night.

As I listen now, I catch the call of a tawny owl somewhere below me. The classic ocarina-like 'howo-huh ho-ho-ho-hoooo' that probably wouldn't have caused panic for ancient humankind, it being familiar and soft in tone. I instinctively reply – much as I do with the ravens. A tawny owl is far easier to stir a response from, but my lips are colder than I expected and the numbness stifles my effort – the result more bovine than owl. I try again but now cannot shape my mouth through laughing at my first attempt. Perhaps I should just enjoy the natural soundtrack.

The brash scream of a barn owl could certainly cause a greater prickle after dark, as perhaps could the 'xylophone trill' call of the tawny. There is some debate as to whether this is a sound made by both sexes or just the male, but a wider acceptance is that it is associated with courtship. It is less often heard than the familiar hoots and 'k-wicks' and has a very different quality. A curious warble that sits somewhere between the throb of a flying saucer from a 1950s B-movie and a Native American war cry from an old Western. It puzzled me when I first heard it, and when it stands alone an owl is not the most obvious source. Waking up cold to it in the middle of the night could send a dream-filled mind into a spiral – the less-uttered sounds from familiar creatures perhaps more likely to cause consternation than the call of the unfamiliar.

I am taking a slow walk back. The tawny owl soon fell silent and the fall into the black tempted me not. Unlike the light of the sun, which penetrates all but the darkest of corners, moonlight can make parts of the night even blacker. Where it shines, so detail is revealed, but the contrast can prevent your eyes from becoming accustomed to the dark itself and it masks the shadows as it does the stars behind it. Tonight, the cloud has thinned and the moon has climbed that little bit higher above the main bulk, so my path, the lane, is glistening silver in response. Where the tarmac remains intact, the light reflects off the tiny little drops of moisture that surround the aggregate chips. The effect is a dappling of moonlight that seems to vibrate. A result, I think, of light diffusion through the soft haze of cloud or my eyes struggling to establish what they see.

I am enjoying the company of the moon. It sits to my right, slightly in my periphery but a steady companion as I make my gentle stroll back. It was a full moon that prompted the slightly unlikely combination of threads that have weaved through this narrative, yet this is the one and only time I have actually walked beneath one. The realisation that the full moon itself was not as symbolic among the Celtic people as I had assumed threatened to trip me up before I had even got going. But instead, it forced me to delve deeper and gain a better knowledge and understanding than I otherwise might. The moon as a timepiece and constant rather than a curio noted for one day in twenty-eight. The night itself more relevant to time's passage than the day, and the lunar influence then layered through every aspect of society. Within such an all-encompassing role, the moon then becomes somewhat incidental. Not so much taken for granted but more an overfamiliarity – it becomes part of our lives rather than an occasional visitor. And as too often happens in such situations, I have not paid it the attention that I might.

Oh! Movement. Something stirred close to the fence line to my right. I didn't see it but heard the sudden shift and rapid pad as it ran up the slope towards the main bulk of the fort. My eyes narrow as I tried to discern a form among the shadows. It surely wasn't there as I walked out, although I have been less stealthy on my return – my mind on the moon rather than what might be around me – so perhaps it had simply tucked up and watched me warily as I passed. The heavier clomp from the second passing of my boots carried greater menace and prompted flight. My first thought is a deer. The Eggardon herd is nearly four dozen strong, but this was a lone animal so perhaps a solitary roe. But the fact it hasn't broken the skyline, and I would definitely make out a deer against the moon-glinted horizon, has turned my thoughts to a smaller animal. There wasn't the lumbering roll of a badger, or the rhythmic 'bom-bom' of a fox. Instead the powerful thuds as it accelerated away has put in mind those long spring-loaded legs of a hare. Perhaps it was 'my' hare – that is a nice thought. Although I probably should explain.

As I was turning left at the Eggardon crossroads after an evening's tench fishing in early summer, a hare loomed in the headlights. It jagged away to the right, into the verge and out of sight, and three leverets followed it. Well, two followed while the third sat unmoved in the glare. I turned the lights off, and then the car engine, waiting until my eyes could determine that it was still sitting tight before flicking the side lights back on. I got out quietly, thinking my bulking presence would provoke a response, but still it sat. It was tiny and fully furred, but hares are precocial, meaning their young are born at an advanced state, fully mobile and with eyes open. I gently nudged it with a finger, beginning to wonder if it was injured, but still it stayed put and I decided that it might be best for me to move it.

I couldn't believe how soft it was, how impossibly delicate, and yet it was warm with a heart thumping fast. I carried it to the point of the verge where I saw the others disappear and then paused for a moment, suddenly aware of the privilege. I had never held a live leveret before and am unlikely to ever hold one again. I was struck in the late of night by the intimacy of the moment.

Then it bit me. Not especially hard, but sharp enough to snap me straight, place it down in the grass and remember that it was a wild animal that didn't much like being held.

Around six weeks later, I made the same late-evening journey and turned the corner to find an almost identical scene. An adult hare quick to flee the glare of the headlights and two leverets swift to follow, while a third loitered a little longer. It had grown considerably between sightings and didn't need my assistance on this occasion, a gangle of legs leading it off behind its siblings. I smiled as I rolled off home, buoyed again by the briefest encounter.

EPILOGUE

THE BLUE MOON

I heard someone whisper 'please adore me', and when I looked, the moon had turned to gold.

(Richard Rodgers and Lorenz Hart, 'Blue Moon')

I REMEMBER THE DAY being warm and the sky broken, and because of what I saw, I am fairly certain it was spring. But the detail is insignificant to the moment itself. The movement, even at such distance, pulled my eyes straight to it. Something high up but falling, and falling vertically and at speed. As I focussed my binoculars, so I made out the broken form and my stomach tightened. Flailing limbs – was it a person? A frozen faller from an aeroplane or a parachutist with a broken ripcord? Time slowed as my mind rattled through the possibilities. Perhaps it was space debris? A prop from a single-engine plane? Or were they fingers splaying at the end of desperate arms?

The object had almost reached the ground when suddenly it split in two, and in that moment my mind replayed the previous couple of seconds with the new information it had received. The gaps filled, and before I had time to cognitively appreciate the fact, I understood exactly what I had seen. Two buzzards, probably at an original height of over 1,000 feet, talons locked and with wings and bodies spinning as they hurtled towards the ground. An avian game of dare that looked perilously close to ending in a dead heat. Each buzzard only a foot or two above the ground as they pressed their wings down hard to break their fall, swinging away from one another in opposite directions as they did so.

I had read before of talon grappling. A behaviour believed to be born through either courtship or territorial dispute. I'd also witnessed moments of it, but they had been fleeting – little more than the insincere handshake shared by footballers before a game. But while it was far from unique, to watch the fall for so long and track it from a speck through to its end was incredibly fortunate. A moment etched into my memory as something singular and spectacular.

It is the unexpected nature of something that tends to knock you sideways. Those snapshots that flash up like photographs decades after the event, which still stir the sense of stupefied awe that struck in the moment itself. The depth swells from the shock of the unforeseen.

We can appreciate things we anticipate, particularly if they surpass expectation. A late-afternoon visit to a winter reedbed or seaside pier to watch the liquid hypnosis of a 10,000-strong starling murmuration – even when you know what's coming you will still leave open-mouthed, and the script changes every day. But there is a different quality when you find yourself stopped in your tracks.

I once journeyed home from work to find a great swathe of starlings in a rippling cloud above our house. I rushed indoors to find Sue already rapt at the window, the flock having been apparently disturbed from their usual roost in the lower Itchen valley and so headed upstream to eventually spend a chattered night in the small clump of cedars in our close. I have had better views of the starling spectacle and having the murmuration swirl directly overhead makes for tricky viewing and a cricked neck. But I barely slept that night due to the excitement, regularly leaning an ear to the window to listen in to the unending racket and reassure myself that the whole thing was real. The unexpected intimacy took an already extraordinary natural event to another level. The memory nestling in with the talon-grappling buzzards and cascade of painted lady butterflies in that select part of my mind that I tend not to access unless I really need to. Like a favourite piece of music or film, the longer that you go without enjoying it, the greater the impact when you do. And by that point, a recollection might stir the same feeling that came at the time, and in turn pull you out of a place you might not want to be.

Of course, you can't possibly be aware of the gravity of these events as they happen. The subconscious is fickle, sometimes dwelling on the unlikely while it disregards the obvious. There will have been a few moments in the past year that might cement themselves into that special corner of memory. The nippy leveret, the melanistic adder, the impossibly perfect young pipistrelle bat (did I even mention him?). But there is one occasion that will definitely wedge itself deep, especially when I ponder it now in the depth of winter. That oozy Eggardon sunset that began this book even though it came midway through the journey. But while the light and sky were special, it was the smell that surprised me most. The intoxicating hang of all those wildflowers mingling with the warmth of the earth. And deeper still came the understanding I had begun to feel. For the people that might have stood where I did and marvelled in the same way. It was around that time that interest began to merge with empathy, a potential tangle of complexities all uncoiling in an instant to make perfect sense.

It seems impossible now, with the temperature dipping sharply and the air cold still. The sunset is just as spectacular, though, the high cloud scattering the pink and orange. But I'm not looking west and I am not even in the place where I meant to be. A quick zip up to the doctor's surgery to collect a prescription offered the chance to reflect. I couldn't simply cut short the span because the solar year just past contained only twelve full moons. There has to be a thirteenth at some point, a Blue Moon as we commonly know it, a shortfall covered in the Coligny calendar by the intercalary months of *Quimonios* and *Sonnocingos*, which slot into the five-solar-year cycle and work much like 29 February does in this age. And the symbolism of a Blue Moon is such that its metaphorical use is familiar to us all, even if its roots seem not to extend further than the early twentieth century. An event very rare in occurrence. Perhaps a politician speaking honestly or an England team winning an Ashes Test match.

To ponder this today, I had intended to stretch that meaning a little further and incorporate the unknown. Between our cottage and the fort on Eggardon sits a pair of stones, marked by Ordnance Survey as a 'burial chamber'. I have mentioned them before and thought that a few words scribbled or recorded while in sight of the site may have resonated. One of the fascinations about the stones is the lack of information about them. It seems accepted that they once formed a cromlech – an arrangement of large stones above a tomb – and that they either fell or were knocked over some considerable time ago. They are partially buried and considerable in size – at least 6 feet long with 3 or 4 feet protruding above the ground – and further stones might be beneath the earth. Whether they mark an entrance to a more significant barrow is speculated but unknown and I rather like the sense of mystery. I have never ventured in to take a closer look and permission would not be difficult to obtain, but I am happy to keep them at arm's length – to maintain the myth.

But then, having collected the prescription, I took a slight detour which took me past a favourite tree. A small, wiry oak that stands alone along a hedgerow, a hole right through its trunk and just a thin frizz of twigs topping the dual spindle of limbs. It has been a familiar friend that we have enjoyed looking out for since we arrived in Dorset. It is the tree I tried to pick out from the eastern ridge, having walked among the house martin sweep, and some years ago Sue drew and painted it in a series of artworks. It recently lost an arm but remains full of character and, as I look at it now, I wonder as I often have as to its age. Its size points to a tree of less than a century, but it is gnarled and wizened from the wind and probably stunted as a result. There is a kink in the trunk, just above the hole, and on the leeward side in a narrow cleft in the bark, a sprout of fern has found root. It is the only green aside from the grass on the verge, but even when in leaf this diminutive tree struggles to splash much colour beyond the grey and brown.

Not that this tree needs any colour to amplify its charm, though, so perfect is it in imperfection. And this early evening, the bone-washed twists are set against an eastern sky of the like I cannot recall. The sunset in the south-west is impressive, yet the moonrise has brought layers of lavender, rose and saffron that are soft like watercolour and yet dazzle like acrylic. And near the top sits the moon, a neat circle of clotted cream topping a slice of rainbow sponge. The view is one that I will not forget. I have no need for a cromlech or prompted thought with an image such as this to sustain me. And I hadn't considered the fact that the moon would be full, even though its coming has been filling newspaper inches and social media posts. The 'Wolf Moon' is causing plenty of interest, although its transatlantic root has prickled my hackles. Perhaps this level of fuss has always stirred and it is only my interest that has piqued, yet I remain stubborn to the Celtic cause. Although that in itself is a thickening of ambiguity. The moon may have structured the lives of the people who once walked these hills, but such was the depth of its influence that the notion of an etymology drawn from a singular passage seems ever more doubtful – particularly given the apparent significance to the Celts of the first quarter.

But although it might seem rather defeatist to conclude that a narrative thread might not have a true beginning or end, the true value runs somewhat deeper. In looking for answers I have found more and more questions, and for as long as there are questions to ask, then there is always an interest to be found. That interest in turn creates a place of continual inspiration and wonder, where I can delve deeply or snatch a quick, five-minute fix, but always find a grounding of self. A balance against the drudge; a restful eddy out of the main flow. Isn't that the place we all seek?

Right now, I'm going to stand and breathe in this scene for a little while longer. It is definitely a Blue Moon moment, even if the moon itself isn't blue. Nor is it a Wolf Moon – not from where I am looking. For now, it can only be the Quiet Moon.

ACKNOWLEDGEMENTS

This book simply wouldn't have happened had it not been for the efforts and support of Jo de Vries and Emma Parkin. Thank you both, and especially Jo for never giving up on it.

Alex Boulton, Laura Perehinec, Simon Wright, Graham Robson, Katie Beard, Andy Lovell and Victoria Hunt have been a pleasure to work with. And thanks to all who have assisted at Flint Books and The History Press.

I reached out to several experts when seeking knowledge and answers regarding the Celts and the Coligny calendar and received so much support. Special thanks to Barry Cunliffe, Andrew Fitzpatrick, Miranda Aldhouse-Green and Caitlin Matthews.

Massive thanks to Dan Kieran and Chris Yates for their wise words and advice, and to Fergus Collins, Will Millard, Verity Sharp, Tim Dee, Hugh Ortega-Breton and Garrett Fallon. My parents, brother and sister and extended family and friends have been relentless in their support and belief – thank you all.

And to Sue, my wife, not just for her support and love but for her beautiful illustrations that are simply perfect.

BIBLIOGRAPHY

Books

Aldhouse-Green, Miranda, *Caesar's Druids: Story of an Ancient Priesthood* (Yale University Press, 2010).

Aretino, Pietro, *The Works of Aretino: Biography: De Sanctis. The Letters. The Sonnets. Appendix* (1926).

Boulton, E.H.B., *A Pocket Book of British Trees* (A&C Black, 1937).

Buczacki, Stefan, *Fauna Britannica* (Hamlyn, 2002).

Caesar, Julius, *Commentarii de Bello Gallico* (58–49 BC).

Camden, William, *Britannia* (1607).

Carlyle, Thomas, *Sartor Resartus* (1836).

Cunliffe, Barry, *The Ancient Celts* (Oxford University Press, 1997).

Darwin, Charles, *The Formation of Vegetable Mould Through the Action of Worms* (1881).

Delamarre, Xavier, *Dictionnaire de la langue gauloise* (Errance, 2018).

Dickens, Charles, *Bleak House* (1853).

Gibbons, Bob, and Peter Brough, *Philip's Guide to Wild Flowers* of *Britain and Northern Europe* (Philip's, 2008).

Goscinny, René, and Albert Uderzo, *Asterix and the Golden Sickle* (Hodder Dargaud, 1960).

Hardey, Jon, et al., *Raptors: A Field Guide for Surveys and Monitoring* (TSO, 2006).
Hegel, Georg Wilhelm Friedrich, *The Philosophy of Right* (1820).
Holding, Michael, *Why We Kneel, How We Rise* (Simon & Schuster UK, 2021).
Lewington, Richard, *Butterflies of Great Britain and Ireland* (BWP, 2003).
McHarg, Ian, *Design with Nature* (Published for the American Museum of Natural History by the Natural History Press, 1969).
Matthews, Caitlin, *The Celtic Tradition* (Element, 1994).
Pliny the Elder, *Natural History* [trans. John Bostock] (2015).
Rousseau, Jean-Jacques, *The Reveries of the Solitary Walker* (1782).
Savignac, Jean-Paul, *Dictionnaire français-gaulois* (Différence, 2014).
Stokoe, W.J., *The Observer's Book of Trees* (Frederick Warne & Co., 1937).
Svensson, Lars, *Collins Bird Guide: The Most Complete Guide to the Birds of Britain and Europe* (HarperCollins, 2009).
Wells, Peter S., *The Barbarians Speak* (Princeton University Press, 2001).
White, Gilbert, *Natural History of Selborne* (Penguin Classics, 1977).
Wood, John George, *Lane and Field* (1879).
Wright, John, *The River Cottage Mushroom Handbook* (Bloomsbury, 2007).

Poems

Bennett, William Cox, 'To a Cricket'.
Brontë, Emily, 'Remembrance'.
Eliot, T.S., 'The Love Song of J. Alfred Prufrock'.
Elliott, Ebenezer, 'Spring'.
Hopkins, Gerard Manley, 'The Windhover'.
Hughes, Ted, 'The Warm and the Cold'.
Pushkin, Alexander, 'Devils'.

Shelley, Percy Bysshe, 'The Sensitive Plant'.
Wordsworth, William, 'To a Butterfly'.

Articles

Arkowitz, Hal, and Scott O. Lilienfeld, 'Lunacy and the Full Moon', *Scientific American*, 1 February 2009 (www.scientificamerican.com/article/lunacy-and-the-full-moon/#).

Barkham, Patrick, 'Hope "Rabbit Hotels" Can Help Britain's Decimated Population Bounce Back', *The Guardian*, 28 November 2021 (www.theguardian.com/environment/2021/nov/28/hope-rabbit-hotels-can-help-britains-decimated-population-bounce-back).

Bartsiokas, Antonis, and Juan-Luis Arsuaga, 'Hibernation in Hominins from Atapuerca, Spain Half a Million Years Ago', *L'Anthropologie* 124.5 (2020).

Brunt, Liam, 'Estimating English Wheat Production in the Industrial Revolution', *Oxford Economic and Social History Working Papers* (1999).

Fitzpatrick, A., 'Night and Day: The Symbolism of Astral Signs on Later Iron Age Anthropomorphic Short Swords', *Proceedings of the Prehistoric Society* 62 (1996), 373–398.

Geddes, Linda, 'The Mood-Altering Power of the Moon', *BBC Future*, 31 July 2019 (www.bbc.com/future/article/20190731-is-the-moon-impacting-your-mood-and-wellbeing).

Harrison, Thomas P., 'Birds in the Moon', *Isis* 45.4 (1954), 323–330.

King, T.J., 'Ant-Hills and Grassland History', *Journal of Biogeography* 8.4 (1981), 329–334.

Lee, Alexander, 'The Great Migration Mystery', *History Today* 70.5 (May 2020).

McKay, Helen T., 'The Coligny Calendar as a Metonic Lunar Calendar', *Etudes Celtiques* 42.1 (2016).

McKie, Robin, 'Early Humans May Have Survived the Harsh Winters by Hibernating', *The Guardian*, 20 December 2020 (www.theguardian.com/science/2020/dec/20/early-humans-may-have-survived-the-harsh-winters-by-hibernating#).

MacNeill, Eóin, 'On the Notation and Chronography of the Calendar of Coligny', *Ériu* 10 (1926), 1–67.

Olmsted, Garrett, 'The Use of Ordinal Numerals on the Gaulish Coligny Calendar', *The Journal of Indo-European Studies* (1988).

Pauly, D., 'Anecdotes and the Shifting Baseline Syndrome of Fisheries', *Trends in Ecology and Evolution* 10.10 (1995).

Racimo, Fernando, et al., 'Archaic Adaptive Introgression in TBX15/WARS2', *Molecular Biology and Evolution* 34 (2017), 509–524.

Raye, Lee, 'Red Kites and Ravens Swooped through Elizabethan London – and Helped Keep the City Clean', *The Conversation*, 25 February 2021 (theconversation.com/red-kites-and-ravens-swooped-through-elizabethan-london-and-helped-keep-the-city-clean-155768#).

Rotton, James and Ivan W. Kelly, 'Much Ado about the Full Moon: A Meta-Analysis of Lunar-Lunacy Research', *Psychological Bulletin* 97.2 (March 1985), 286–306.

Russell, Miles, et al., 'The Girl with the Chariot Medallion: A Well-Furnished, Late Iron Age Durotrigian Burial from Langton Herring, Dorset', *Archaeological Journal* 176.2 (2019), 196–230.

Sánchez-Bayo, Francisco, and Kris A.G. Wyckhuys, 'Worldwide Decline of the Entomofauna: A Review of Its Drivers', *Biological Conservation* 232 (2019), 8–27.

Stendall, J.A.S., 'Observation on the Woodcock Carrying Young', *Irish Naturalists' Journal* 1.5 (1926), 89–91.

Swift, Cathy, 'Celts, Romans and the Coligny Calendar', *Theoretical Roman Archaeology Journal* (2002).

Reports

Butterfly Conservation, *The State of the UK's Butterflies* (2015).

State of Nature Report (2013).